Paddington Station

Paddington Station

Its history and architecture

Steven Brindle

Second edition

ENGLISH HERITAGE

Published by English Heritage, The Engine House, Fire Fly Avenue, Swindon SN2 2EH
www.english-heritage.org.uk
English Heritage is the Government's lead body for the historic environment.

First edition published 2004
Reprinted 2009
Second edition 2013

ISBN 9 78184 8020894

Product code 51682

British Library Cataloguing in Publication data
A CIP catalogue record for this book is available from the British Library.

For more information about images from the English Heritage Archive, contact Archive Services Team, The Engine House, Fire Fly Avenue, Swindon SN2 2EH; telephone (01793) 414600.

Brought to publication by Rachel Howard, Publishing, English Heritage.

Typeset in Charter 9.5 on 11.75

Edited by Susan Kelleher
Indexed by Ann Hudson
Page layout by Hybert Design

Printed in the UK by Butler Tanner and Dennis Ltd.

Front cover
Platform 1, with the clock over the original departure side entrance. [© English Heritage AA061959]
Frontispiece
Elevated interior view, taken in 2003. [© English Heritage K030818]
Back Cover
A late 19th-century carriage-door panel with the arms of the GWR.
[STEAM – Museum of the Great Western Railway, Swindon]

CONTENTS

Preface and Acknowledgements vi

1 Brunel and the Great Western Railway 1

2 The First London Terminus, 1835–1850 12

3 Building the New Station, 1850–1855 26

4 Maturity, 1855–1947 51

5 The Station at Work, 1855–1947 70

6 Decline, Revival and the Future, from 1948 90

7 The Architecture of the Station 99

8 Related Buildings and Structures 128

Notes 158

Bibliography 173

Index 177

PREFACE AND ACKNOWLEDGEMENTS

I am going to design, in a great hurry, and I believe to build, a Station after my own fancy;
that is, with engineering roofs etc. It is at Paddington, in a cutting,
and admitting of no exterior, all interior and all roofed in ...

Isambard Kingdom Brunel, in a letter to Matthew Digby Wyatt, 13 January 1851

The best way to approach Paddington is from its underground station. You struggle through the crowded, low-ceilinged anonymity of the subways, through a ticket barrier: stairs and an escalator loom ahead. Suddenly the ceiling ends and there is an arched roof high above, and as you ascend that happily placed staircase the whole tremendous interior reveals itself at once. You are standing in the middle of the broad central span with, on a fine day, light flooding in from the Lawn roof behind and, almost 215 metres away, light silhouetting the delicate tracery at the far end. The milling crowds, the long perspective, the waiting trains and the light in the distance produce an atmosphere of expectation: the experience of travel is heightened here in a way that few other British stations can match. It is a huge space, but somehow it never overwhelms; it is dark enough to have mystery, but at the same time you can understand at once where you are and where everything is. The proportions are perfect, with the inevitability of a great work of art. It is one of the great interiors of London.

Paddington, the Great Western Railway (GWR) and their principal designer, Isambard Kingdom Brunel, are at the heart of railway history. Railways were one of the defining phenomena of Victorian Britain and one of Britain's many great gifts to the world, so the story of the building of the GWR has a place in world history. This is why Paddington, together with the other principal elements of the main line to Bristol, has been a candidate for inscription as a World Heritage Site.[1] The name and memory of the GWR continue, however, to have a resonance which goes well beyond these matters of historical fact. Why? The company built a superb railway network, though other companies did too. It provided the west of England and Wales with excellent train services, though they were not always so. It had a remarkable record of engineering excellence, though in some respects, in matters as important as electrification, it was rather slow. It was a reliable and safe investment, usually yielding around 4 or 5 per cent, year after year, to thousands of investors.

In the 19th century the GWR was often controversial and occasionally unpopular, but in its golden age from *c* 1875 to 1914 and its silver age between 1923 and 1939, the company attracted a remarkable degree of loyalty and affection from its servants (as they were always called), passengers and shareholders. The GWR achieved tremendous things, while maintaining a reputation for absolute integrity, honesty and competence, for over a century. There are very few other vanished companies whose 'brand' remains so strong, and the GWR ceased to exist over 55 years ago.

So, in writing a history of Paddington Station, one is uncomfortably aware that there is a great deal to live up to. The station has meant so much to so many people, and it embodies so many stories, that there is a certain presumptuousness about the whole business (especially if, like the present author, you come from Lancashire, and not from GWR country at all). I am therefore deeply grateful to the many people who were so generous with their time and knowledge during the writing of the first edition of this book. I am particularly grateful to Robert Thornton, senior architect at Network Rail plc for all his help and advice; to Ian Nulty and Mervyn Hellen, formerly the records manager and senior conservator at Network Rail's Western Region Plans Room in Swindon, for all their help and kindness on numerous visits; to Tim Bryan, former director and Felicity Ball, curator, at STEAM, the museum of the Great Western Railway at Swindon; to Peter Rance of the Great Western Trust and Peter Treloar of the Great Western Society; to Alan Garner and John Spellar of the Broad Gauge Society; to Michael Richardson, keeper of special collections at Bristol University Library, to Ed Bartholomew of the National Railway Museum; to Graham King of Westminster City Council; to Richard Brown of Oxford Archaeology; and to Rob Naybour of Weston Williamson. I have been much helped by the staff of the British Library, the Hulton Getty Picture Library, the library of the Institution of Civil Engineers, the National Portrait Gallery, the National Archives, the Royal Institute of British Architects Drawings Collection, the print room of the Victoria and Albert Museum, and Westminster Archives, and would like to record my grateful thanks to them all.

Various other friends read drafts of the text for the first edition and provided help and advice. I am especially grateful to Alan Johnson, Andrew Saint, Gavin Stamp and James Sutherland. At English Heritage, I am grateful to Philip Davies for proposing that the research might be turned into a book, to my then managers, John Thorneycroft and Paul Velluet for their support, and to Tim Jones, Keith Falconer, Val Horsler, Cathy Philpots and Robin Taylor for their help. Thanks are due above all to English Heritage's publications team, led by the editors, Rowan Whimster and René Rodgers, for turning my draft into a book.

Work for this second edition of this book has incurred several more debts of gratitude: many of the people listed above have helped again. Dave Keeley and Jay Carver of Crossrail Ltd; Robert Thorne and Richard Pollard of Alan Baxter Associates; and Nick Hartnell, station manager at Paddington, have been particularly kind. Nigel Corrie, who took the excellent new photographs for the first edition, has added to them for the second, and Mark Fenton has very kindly updated Philip Sinton's graphics for it. I would particularly like to thank Rachel Howard, Susan Kelleher and Linda Elliott of Hybert Design, for all their work to bring the second edition to publication. We are very grateful to Network Rail plc, who own and run Paddington, for all their help in the writing and production of this book.

The sources for Paddington's history are vast and complex, and much has never yet been touched by historians, and much that might have gone into this book has been left out for one reason and another. Any mistakes or omissions are the author's sole responsibility, and I would be grateful for any suggestions on these scores. Lastly, this revised edition is dedicated to my mother, with love.

Brunel and the Great Western Railway

To understand Paddington Station you need to understand the Great Western Railway. Isambard Kingdom Brunel, the Great Western's engineer, called it 'the finest work in England' (Fig 1.1).[1] Many people since have regarded Brunel's line from London to Bristol as a masterpiece of early railway engineering and the GWR as a whole has a special place in railway history. Yet today, when engineering is a mature profession, earth moving and construction are mechanised and easy, and railways are rather taken for granted, it is difficult to grasp the physical and intellectual scale of the first railway builders' achievements.

In 1824 a group of Bristol merchants assembled to make the first attempt to found a company to build a railway linking Bristol with London, 110 miles (177km) to the east. It was unsuccessful, as were the next five.[2] However, Bristol needed a railway. In the 18th century it had been the busiest port in England after London and the hub of the Atlantic trade, but by the 1820s it had clearly been overtaken by Liverpool. Bristol suffered from its cramped harbour frontage on the tidal River Avon: they had responded vigorously with the remarkable Floating Harbour project (1809) which had dammed the original course of the river to form a huge enclosed dock while diverting the flow of the river down the 'New Cut' to the south. This helped, as had the opening of the Kennet & Avon Canal in 1810, but the Bristolians were still on the back foot. The opening of the Liverpool & Manchester Railway in 1830 was a further blow, and to make matters worse, the Liverpudlians were already actively planning further lines: the Grand Junction Railway to link Warrington to Birmingham, and the London & Birmingham Railway (L&BR), giving them a trunk route all the way to the capital. Matching the northern rivals with a railway to London seemed to be Bristol's only hope.[3]

In the autumn of 1832 another group of four merchants met in an office on the site of Temple Meads Station and formed a committee to build 'the Bristol Railway'. This time the venture took off. A company was formed, with representatives of Bristol's four corporate bodies: the City Corporation, the Bristol Dock Company, the Company of Merchant Venturers and the Chamber of Commerce. The Bristol & Gloucestershire Railway, a rather ambitiously titled 10 mile (16.1km)-tramway for horse-drawn coal wagons, was also represented.[4] On 21 January 1833 they held their first public meeting and in February they raised funds for a survey. The job was offered to whoever would carry out the survey most cheaply and the company further stated that they would appoint whoever submitted the cheapest building estimate as their engineer.[5] There were already some local candidates in the picture, including William Townsend, the Bristol & Gloucestershire Railway's surveyor. Two others, William Brunton and Henry Price, had in May 1832 published a speculative proposal on their own initiative for a railway to London, taking a rather hilly route via Bath, Trowbridge, Devizes, Hungerford and Reading.[6] One of the directors, Nicholas Roch, however, put forward the name of a 27-year-old London engineer: Isambard Kingdom Brunel (1806–59), son of the celebrated French-born engineer Sir Marc Brunel (1769–1849).[7] The younger Brunel was already well known in Bristol, having won a competition for a bridge to span the Avon Gorge at Clifton and served as the engineer to the Bristol Dock Company.

In 1833 the younger Brunel had never yet worked on a railway, but then very few people had: the only substantial railways using steam locomotives were the Stockton & Darlington (1824), the Canterbury & Whitstable (1825) and the Liverpool & Manchester (1830) (Fig 1.2).[8] In 1831 Brunel had travelled to the north of England looking for work; he visited the Stockton & Darlington Railway and on 5 December 1831 he took a journey on the Liverpool & Manchester Railway. This seems to have been the total extent of his experience of railways when he set about surveying and designing the Great Western. He had the benefit, however, of

a peerless education in engineering, thanks to his brilliant father. In 1820 he went to school in Paris and became a pupil to a celebrated engineer and clockmaker, Louis Breguet. From the age of 16 to 19 he helped his father, carried out surveys and served an apprenticeship with the mechanical engineering firm of Maudslay, Son & Field.[9]

In 1825 Marc Brunel was appointed engineer to the Thames Tunnel Company and began work on that daring undertaking, the first tunnel ever driven beneath a significant body of water, using the 'tunnelling shield' which he had patented in 1818. From 1827 his 21-year-old son was the resident engineer on the project, keeping a horde of unruly labourers in check and struggling to keep the Thames at bay. The project was perilous in the extreme: the tunnel workings flooded in May 1827, and again on 11 January 1828. Six men were drowned, and Isambard himself was seriously injured. Work on the tunnel came to a halt. It was while convalescing from this accident that Isambard first went to Bristol. On 18 March 1831 he won the second competition for the proposed Clifton suspension bridge across the Avon Gorge, but in the event the Clifton Bridge Committee could not raise the money to start

Figure 1.2
The Liverpool Road Station, Manchester – the simple façade of the office and booking hall, looking much like a terraced house, belied its status as the world's first railway terminus.
[Courtesy of Manchester Libraries, Information and Archives, Manchester City Council]

work. However, he had made valuable contacts in Bristol and in 1832 he was appointed as engineer to the city's Dock Company. The Floating Harbour suffered from a serious silting problem and Brunel designed a number of works, including two dredgers, to tackle this.[10]

For Brunel, the chance to survey the new Bristol Railway was the greatest professional opportunity he had yet had. Nevertheless, he dealt very frankly with the directors, criticising their emphasis on cheapness: 'You are simply giving a premium to the man who will make the most flattering promises. The route I will survey will not be the cheapest – but it will be the best.' Many would have found such an attitude off-putting, but on 7 March 1833 the directors voted by a majority of one to appoint Brunel to make their survey in partnership with William Townsend.[11]

On 9 March 1833 Brunel and Townsend set out from Bristol to begin work on their survey. It was clear from the outset that this was no equal partnership. Townsend was gratifyingly deferential: he was happy to remain in Bristol, responsible for that end of the line, and to work to Brunel's instructions. Brunel made an initial survey of the whole vast area from horseback, hired local surveyors to assist him, opened discussions with landowners and produced the initial plans. It was arduous work. Brunel's son was to write:

His diary of this date shows that when he halted at an inn for the night but little time was spent in rest, and that often he sat up writing letters and reports until it was almost time for his horse to come round to take him on the day's work. 'Between ourselves', he wrote to Hammond, his assistant, 'it is harder work than I like. I am rarely much under twenty hours a day at it.'[12]

In this way, he finished the survey in 10 weeks flat. On 30 July 1833 Brunel explained his proposals at the Bristol Railway's first public meeting. He estimated that the line would be 116 miles (196km) long and cost £2.5 million, complete with stations and rolling stock (it turned out to cost two and a half times as much).[13]

A London committee was formed and the first joint meeting of the two committees was held on 19 August, at which the company renamed itself the Great Western Railway.[14] At the same meeting Charles Alexander Saunders was appointed secretary to the London committee: he turned out to be one of the key figures in the creation of the railway and one

Figure 1.3
Charles Saunders, the first company secretary of the GWR.
[© National Railway Museum/Science and Society Picture Library]

of Brunel's staunchest supporters (Fig 1.3). The speed and efficiency with which Brunel had carried out the survey had impressed the new board and he was appointed as the company's engineer on 27 August.[15] His salary was £2,000 plus £300 in lieu of travelling expenses. This looked princely, but Brunel was expected to find 'all personal charges for his general professional duties' from this, including the salaries for his assistants and draughtsmen. He was expected to devote himself to 'the general service of the company' and was naturally barred from working for any other railway without the permission of the board: in the event he was to be the chief engineer for several of the GWR's allied lines, such as the Bristol & Exeter and the South Devon Railway.[16]

With a whole railway to design, Brunel swiftly rented a London house at 53 Parliament Street and hired several assistants, though it was a while before he established a settled office. He commissioned a specially constructed *britschka* (a large four-horse coach), with sleeping accommodation. Brunel lived mostly on the road for several months, turning his survey into a full design, treating with landowners, writing all the correspondence, checking all the accounts and joining in the work of selling shares. By the end of the year, just nine months after Brunel had ridden out of Bristol with William Townsend, complete plans for the 116 miles (196km) of railway were deposited with the Parliamentary Bills Office. Apart from the London end, where Brunel proposed a terminus near Vauxhall Bridge, the line was shown mostly as built. Brunel proposed a tunnel beneath Sonning Hill, Berkshire, where in fact a cutting was used, and between Bristol and Bath he had the line crossing the Avon three times: as built, it just crosses the river once.[17]

Raising the immense sums of money needed proved very difficult. As an interim measure the first bill was only for the construction of the outer sections of line, from Bristol to Bath and from Reading to London: opponents derided it as 'neither Great, nor Western, nor even a Railway'. The bill went through its committee stage in April 1834. Brunel was the most important witness and performed superbly during several days of questioning. The bill, however, was thrown out by the House of Lords in July on the grounds that the company's financial security was inadequate. A new prospectus was issued, more shares were sold and a revised bill was entered in February 1835. Brunel was questioned for 11 of the 40 days that the bill was before Committee and this time they were successful. The Great Western Railway Bill received royal assent in August 1835: it had cost the company £88,710.[18]

Brunel and the infant GWR stood apart from the other railway companies. Railways began in the north of England, based on the experience and the technologies that had developed around coal mines, coal tramways and canals. Most of the early railway engineers were northerners, and many, like George Stephenson, were 'practical men', with little formal education but rich in experience. Most of the early railway entrepreneurs and financiers were northerners too, above all from Liverpool and Manchester. It is remarkable how little of the impetus for the first railways came from London.[19] The GWR was set apart by being a Bristolian venture: Bristol was probably the only southern English city other than London which would have had the means and the confidence to do something on this scale. Brunel was set apart by being a Londoner, half-French and well educated.[20]

In planning the Great Western, Brunel acted with remarkable boldness and originality. His rivals Brunton and Price had proposed a southern route, taking in as many significant towns as possible, including Bath, Bradford-on-Avon, Trowbridge, Devizes, Hungerford and Reading. This would have delivered the traffic of these towns, but at the cost of a winding route with steeper gradients, inevitably slowing everything down. Brunel instead chose a northern route, swinging north from Bath and across broad, level, relatively empty country from Chippenham to Didcot. From there his line followed the Thames valley south to Reading and east to London (Fig 1.4). Brunel's line is remarkably straight: none of its curves have a radius of less than a mile. It is also remarkably level: the first 50 miles (80km) from London to Didcot rise at a gradient of 1 in 1320 (4ft per mile; 1.3m per km) and the rest is nowhere steeper than 1 in 660, except for two inclines of 1 in 100 at Wootton Bassett and the Box Tunnel.

Brunel conceived of the Great Western Railway as a complete system, 'the best of all possible railways' with all its interdependent parts, rolling stock and equipment designed together and executed to the highest standards.[21] His vision was of a trunk line from London to Bristol, level and straight and therefore fast. The branch lines could come later: what mattered was choosing the best route to provide the trunk for those branches. By comparison with Brunel's strategic vision, Brunton and Price's plan was a short-sighted join-the-dots exercise, but Brunel was nevertheless criticised for taking his line through so much empty country. To survey and plan so level a line over such great distances, working from horseback, using the simple maps and surveying equipment then available was in itself a remarkable achievement. To do this, and then design the whole railway in just nine months, was extraordinary.

Brunel's track design was equally original, departing radically from the 'Stephensonian' model. George Stephenson modelled his railways on the horse-drawn tramways of the North-East where he grew up. Most of them had a gauge (that is, the breadth between the rails) of 4ft 8½in. (1.44m), and this is what Stephenson adopted for both the Stockton & Darlington and the Liverpool & Manchester Railways.[22] In 1833 the London & Birmingham Railway was begun with his son Robert Stephenson as engineer. In the same year work began on the Grand Junction Railway to link Birmingham to the Liverpool & Manchester Railway: this too adopted the 4ft 8½in. gauge. These Stephensonian lines were originally laid on stone blocks, which had a tendency to settle unevenly under the weight of the trains.

Brunel proposed a different system. His rails would be laid on longitudinal wooden beams, held down on a bed of tightly packed sand by being fixed to wooden piles driven into the ground.[23] Moreover, they would be 7ft (2.14m) apart. Brunel favoured a broad gauge for

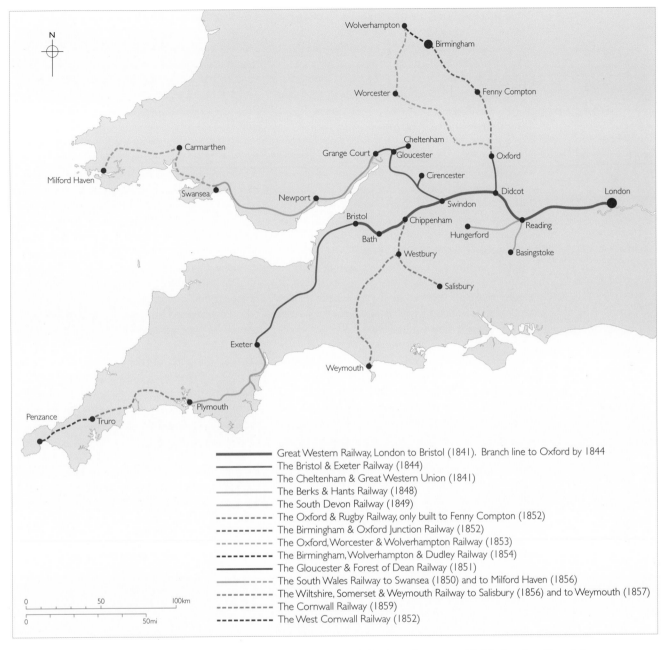

Great Western Railway, London to Bristol (1841). Branch line to Oxford by 1844
The Bristol & Exeter Railway (1844)
The Cheltenham & Great Western Union (1841)
The Berks & Hants Railway (1848)
The South Devon Railway (1849)
The Oxford & Rugby Railway, only built to Fenny Compton (1852)
The Birmingham & Oxford Junction Railway (1852)
The Oxford, Worcester & Wolverhampton Railway (1853)
The Birmingham, Wolverhampton & Dudley Railway (1854)
The Gloucester & Forest of Dean Railway (1851)
The South Wales Railway to Swansea (1850) and to Milford Haven (1856)
The Wiltshire, Somerset & Weymouth Railway to Salisbury (1856) and to Weymouth (1857)
The Cornwall Railway (1859)
The West Cornwall Railway (1852)

several reasons. He believed that vehicles with a wider wheelbase would be more stable, and thus safer, especially at high speeds. It would also allow for larger wheels, which would reduce friction with the rail and within the engine, achieving smoother running. A larger chassis could also allow for larger locomotives and rolling stock. There seems little doubt that the broad gauge made the Great Western more expensive: the lines had a 'maximum width of passage' of 13ft (3.96m) as against 10ft (3.05m) on the London & Birmingham, so all the bridges, tunnels and embankments had to be bigger.[24] Brunel believed, though, that the

combination of the broad gauge with his level and relatively straight track would make the GWR's trains altogether faster.[25] The Stephensonian 4ft 8½in. gauge was already established, however, as the industry standard.

Brunel had not decided to rebel against the 4ft 8½in. gauge from the outset: the first Great Western Railway Bill of 1834, like most railway bills, had specified use of the standard gauge; had it passed, the broad gauge would probably never have existed. Brunel changed his mind between then and the passage of the second bill in February 1835.[26] He persuaded Lord Shaftesbury, who was overseeing the drafting of the

Figure 1.4

The Great Western Railway – the first London to Bristol route designed by Brunel in 1835 and the main routes that were open or under construction (shown as broken lines) by 1859.
[© English Heritage]

5

bill, to omit the clause about gauge, leaving the GWR free to determine the matter for itself. On 15 September 1835, he wrote a long letter to the GWR's board of directors, setting out his arguments for a broader gauge in some detail. He considered the arguments against and said that, in his view, the only convincing one was the inconvenience that could arise in making a junction with the London & Birmingham Railway (as the company was considering having a joint London terminus with the London & Birmingham Railway at Euston, as discussed in the next chapter).[27]

The directors agreed to his proposal on 29 October 1835, though this was not publicly announced until August 1836.[28] A protracted public debate on the 'break of gauge' ensued during the construction of the line. Representatives of the northern railway establishment had bought shares in the GWR and secured places on its board. This group, known as the 'Liverpool party', lobbied against Brunel and the broad gauge. In July 1838 the GWR commissioned outside experts to advise them: they too were hostile. The first section of the track, laid on the west side of London using Brunel's way-beams and piles, turned out to be over-fixed: the piles had no 'give' in them, forming 'bumps' under the axles, and the passengers had a very hard ride as a result. The board decided that the piles would have to be removed and the track re-laid, at great extra cost. Worse still, the

GWR's engines performed poorly in their first trials and between October and December of 1838 Brunel's career and the future of the broad gauge hung by a thread. Brunel and Daniel Gooch, his young locomotive superintendent, did not believe that the speed trials had been fair. They made adjustments to the *North Star* and this seemed to do the trick: further speed trials on 27 December were startlingly successful (Fig 1.5). The board believed that Brunel was vindicated and on 7 January 1839 a special meeting of the company voted to continue with his design by a majority of 7,792 to 6,145.[29]

In 1835 Brunel moved to 18 Duke Street, Westminster, where he set up an office under Joseph Bennett, his chief clerk, who stayed until Brunel's death in 1859 (Fig 1.6). Brunel seems to have borne the office costs of Duke Street out of his £2,000 salary, but in September 1835 the board authorised him to appoint three resident engineers 'at salaries not exceeding £600' to supervise work on the line.[30] By the time it was nearing completion, their number had risen to seven: G E Frere, George T Clark and T E Marsh for the western section; Thomas A Bertram, John W Hammond and Robert P Brereton for the eastern section; and William Glennie for the Box Tunnel.[31]

Brunel was a difficult man to work for. He was loyal to reliable subordinates but he treated under-performers with great harshness, on occasion culminating in their abrupt

Figure 1.5
The North Star, *the GWR's first reliable locomotive.*
[Great Western Trust Collection]

Figure 1.6
Brunel's office at 18 Duke
Street, Westminster,
probably photographed
shortly after his death
in 1859.
[Elton Engineering Books]

dismissal.[32] He was not a good delegator. One of the great puzzles about the building of the GWR is the extent to which Brunel was responsible for designing it all, for he seems to have been very jealous of his primacy in this area and it is difficult to understand how far he shared this responsibility. Brunel certainly laid out the route and designed the track, and his sketchbooks contain great numbers of drawings for bridges, cuttings, tunnels, stations, cottages, sheds and machinery of every kind.[33] These seem to have gone to assistant engineers, or occasionally to collaborating architects (as at Bristol Temple Meads), to be worked up: Brunel seems to have maintained a close interest in almost everything. Designs would be sent to the drawing office at Duke Street, revised, corrected. Eventually, superb finished drawings would be produced, coloured in watercolour. Great numbers of them survive in Network Rail's archives.[34] The originals all evidently remained with Brunel, later passing to the GWR. There is no evidence of Brunel providing the contractors with copies, and it seems more likely that they were obliged to send draughts-

men to Duke Street to make tracings for their own use, though little or no trace of these survives.

Brunel was involved at every stage: he was notoriously unable to leave detail to others.[35] He handled most of the detailed negotiations with difficult landowners, on occasion varying the designs to suit their demands.[36] He wrote the draft contracts for every job and proof-read all the final contract documents, frequently noticing errors. He let the contracts and he supervised the contractors. The obverse of Brunel's olympian self-confidence and decisiveness was an unwillingness to recognise others as equals. The one famous exception, by his own statement, was his rival Robert Stephenson, engineer to the rival London & Birmingham and champion of the 4ft 8½in. gauge.

In retrospect, Brunel's towering reputation is such that his appointment and his triumphant completion of the railway seem inevitable. At the time, he was a controversial figure, with many detractors. In 1887 a Bristol historian, John Latimer, summed up the view of those who thought that Brunel had been altogether too careless with Bristol people's money, lamenting 'the deplorable error of the original board in neglecting the sober-minded, practical and economical engineers of the North … and in preferring to them an inexperienced theorist, enamoured of novelty, prone to seek for difficulties rather than to evade them, and utterly indifferent as to the outlay which his recklessness entailed upon his employers'.[37]

This was not really just: Brunel had constantly to keep economy in mind as it soon became apparent that his initial estimate of £2.5 million was far too low. He went to extraordinary lengths to get value for money out of his contractors, becoming famous, or notorious, for what F R Conder called 'an extreme and unprecedented insistence on excellence of work'. GWR specifications were unusually strict and Brunel insisted on them being carried out to the letter, often delaying payment and sometimes withholding it altogether. One result was that the contractors learnt to push their prices up, another was that on occasion it proved difficult to let the contracts.[38] Two of the contractors, William Ranger and David McIntosh, brought major lawsuits against the GWR, which were not resolved until 1859 and 1865 respectively.[39]

The work was on a huge scale, involving thousands of labourers. It is sobering to reflect that the great embankments, cuttings and tunnels, for which millions of cubic yards of earth and rock had to be moved, were all dug by hand by navvies, who were normally only equipped with spades, picks, shovels and wheelbarrows,

Figure 1.7
An early GWR train crosses Brunel's Maidenhead Bridge in J M W Turner's painting Rain, Steam, and Speed – The Great Western Railway, c 1844.
[© The National Gallery, London]

assisted by hundreds of horses.[40] Such scenes of back-breaking labour were shortly to be seen all over Britain as 'railway mania' took off.

From 1836 to 1838 the line grew westwards from London on immense embankments. Two major obstacles had to be crossed: the River Brent at Hanwell just west of Ealing and the River Thames at Maidenhead. Brunel designed superb crossings for both. That over the Brent, later named the Wharncliffe Viaduct, was the place where work commenced on the GWR, in November 1835.[41] From there the track led west on broad embankments to Maidenhead. There, to span the 100yd (91.4m)-wide Thames, Brunel designed a bridge with two immense brick arches 128ft (91m) wide but with a rise of only 24ft 3in. (7.5m) high, the widest arches that had ever been built (Fig 1.7).[42] The line opened between Paddington and Maidenhead on 31 May 1838, in circumstances of chaotic haste.[43] Further stages opened, reaching Twyford in July 1838, Reading in March 1840, crossing the Thames valley to Steventon near Didcot in June 1840, and to the Faringdon road in the Vale of the White Horse in July.[44] Meanwhile, the line had crept up the Avon valley from Bristol to Bath. This was more difficult terrain and this relatively short stretch only opened in August 1840: Bristol Temple Meads Station was still under construction.[45]

This left a gap to the east of Bath, the most difficult part of the whole line: Box Hill rose in the path of the railway, at the point where the limestone ridge drops into the steep-sided Avon valley. Of the 13 miles (20.9km) of track between Bath and Chippenham, scarcely one mile of it is within 10ft (3m) of the natural ground level: it all had to be sunk in cuttings or raised on embankments. The most difficult task of all was digging the Box Tunnel, 1⅞ miles (3.0km) long through solid rock with another mile through deep cuttings to the east. The construction of the tunnel was an appalling epic, lasting five years, with three main contractors in succession. Six shafts had first to be sunk to the level of the line, 30ft (9.1m) in diameter and ranging from 70ft (21.3m) to 300ft (91.4m) deep: the tunnel was then dug in each direction, from the foot of each shaft. A ton of gunpowder and a ton of candles a week were consumed, as a thousand or more navvies hacked their way through the rock. At the beginning of 1841 Brunel persuaded the contractor George Burge to raise his workforce from 1,200 to 4,000. At last, in April 1841, the

Figure 1.8
The Box Tunnel – a
lithograph by J C Bourne
from his History and
Description of the Great
Western Railway. *Brunel*
designed the great portal
himself.
[STEAM – Museum of the
Great Western Railway,
Swindon]

Box Tunnel was finished (Fig 1.8). It is said to have cost the lives of over a hundred men.[46] In June the GWR's main line was complete from London to Bristol, at a cost of £6,150,000.[47]

In fact the Great Western's trains could already run a good deal further, to Bridgwater in Somerset. In 1836 the Bristol & Exeter Railway had been established. Another Bristolian venture, it had many shareholders and two directors in common with the GWR. It also appointed Brunel as its engineer and was similarly laid to the broad gauge. The Bristol & Exeter company built a second station at Bristol Temple Meads, close by the GWR. Now, trains could either run into the GWR's original terminus, or through the Bristol & Exeter station on to the South-West. The Bristol & Exeter opened as far as Bridgwater on 14 June 1841 and to Exeter in 1844.[48]

In their wider context, the building of the GWR and the battle for the broad gauge could be seen as almost the last chapter in the contest between Bristol and Liverpool for dominance of the Atlantic trade. Early in 1836, just as work was beginning on the western end of the line, a group of GWR directors and investors set up the Great Western Steamship Company, again with Brunel as engineer. He designed the SS *Great Western* for them: built in Bristol, launched in 1838 and fitted out in London, it was the first transatlantic steamship and an absolute success. The Liverpudlians responded with the SS *Sirius*, a hastily adapted steamer from the Dublin run and a much inferior vessel, and for a while it seemed that Bristol had regained the lead; Jack Simmons has observed how the Liverpool Party's campaign to oust Brunel and gain control of the GWR reached its height around the time that the *Great Western* was being completed.[49] The Great Western Steamship Company next commissioned Brunel to design the world's first ocean-going iron steamship, the SS *Great Britain*: part-way through construction he made its design even more revolutionary by adopting the newly invented propeller as its principal means of propulsion. This superb vessel, also built in Bristol's Floating Harbour, was launched in 1843 and made its maiden voyage to New York in 1845.[50] When Prince Albert set off from Paddington to attend its launch, it really seemed as if Brunel's vision of a high-speed transport network from London to the Americas was about to become a reality. Paddington would have been the London terminus for New York. What let Bristol down was the smallness of its docks, 6 miles (9.7km) up the Avon, compounded by the sloth of its Dock Company and the complacency of its merchant class in comparison with Liverpool's entrepreneurial spirit. By 1850 the railway was complete and a triumphant success, but the battle for the Atlantic was lost. For the rest of the 19th century the big transatlantic steamships, including the *Great Britain* herself, sailed from the River Mersey.[51]

In the railway mania of the 1840s, new lines were being planned and built all over the country in circumstances of feverish competition. Several new lines were proposed for tactical reasons, to 'spoil' a proposal from a rival company. The GWR was one of the three leading companies, in competition with the two great standard-gauge empires, the London & North Western (formed by the merger of the London & Birmingham and the Grand Junction Railway in 1845) and the Midland Railway (also formed in 1845, by the merging of four companies). Like these two, the GWR was building up a network of smaller companies under its influence. In 1844–5, the GWR supported the establishment of, and bought large shareholdings in the South Devon Railway; the South Wales Railway; the Cornwall Railway; the Wiltshire, Somerset & Weymouth Railway; the Oxford, Worcester & Wolverhampton Railway; the Berkshire & Hampshire Railway; the Oxford & Rugby Railway; and the Monmouth & Hereford Railway.[52]

These GWR satellite companies laid their tracks to the broad gauge. In the 1840s the 'gauge war' broke out between the GWR and the big standard-gauge companies. The GWR fought hard, running broad-gauge tracks on its own lines and those of its satellite companies all over the South-West, the West Midlands and South Wales – for example to Exeter (1844), Oxford (1844), Gloucester (1845), Cheltenham (1847), Plymouth (1849), Birmingham (1852), Swansea (1852) and Wolverhampton (1854). The gauge war became the subject of national controversy and long parliamentary battles. It is a very complex story. It must suffice here to say that, though the broad gauge's technical

excellence was widely acknowledged swift developments in locomotive and track design improved the performance of standard-gauge trains, eroding the advantages of the broad gauge.[53] In the long run the broad gauge's technical merits were outweighed by the far greater mileage of standard-gauge lines that had been built and by the obvious inconvenience of having two different railway gauges in one country.

The Gauge Act of 1846 recognised 4ft 8½in. as the national standard; thereafter the GWR found it very difficult to secure permission for new lines with the broad gauge. In order to expand northwards the company was obliged to buy or build standard-gauge and mixed-gauge lines itself. As a result the first mixed-gauge track was run into Paddington, the citadel of the broad gauge, as early as 1861 (Fig 1.9). The GWR gradually converted the outlying reaches of its network to the standard gauge but the men who ran it were loath to abandon the broad gauge. At last in 1892 the company converted its remaining tracks over a weekend of gigantic and minutely organised effort.[54]

As the GWR's engineer, Brunel did not see himself as simply designing and building a fixed system of track and rails. From the outset he envisaged it as a functioning whole, designing his track with its broad gauge and generous curves to allow for smoother running and higher speeds. The stations were seen in the same way: his sketchbooks make it clear that Brunel started his station designs with track layouts to plan the movement of rolling stock, fitting the movement of passengers and staff around that, and in turn fitting the buildings around these interlocking functions.

Brunel also felt fully responsible for the locomotives and rolling stock, but he was not really a mechanical engineer and it was here that his commanding omniscience ran out. In 1836–7 he ordered the GWR's first locomotives from the Vulcan Foundry at Warrington, the Haigh Foundry near Wigan, and Mather, Dixon & Co in Liverpool, to be delivered by canal to Paddington (it is interesting that all the manufacturers were in the North-West, within a short distance of Liverpool).[55] Brunel laid down specifications for a number of engine designs, all very idiosyncratic and unlike any others, with small boilers to keep the overall weight down and very large driving wheels. The specification was so odd as to render it impossible for the makers to build satisfactory

Figure 1.9
A late 19th-century view of the Paddington train sheds. Note the mixed gauge with three rails, dating this photograph to 1892 or before.
[City of Westminster Archives Centre]

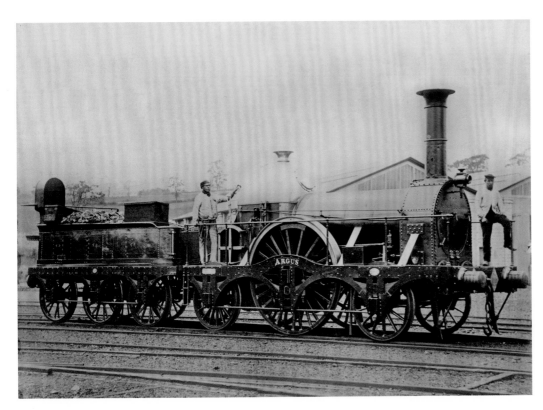

Figure 1.10
The Argus, *built in 1842,*
was 1 of 62 locomotives of
Gooch's Firefly *class, with*
7-foot diameter driving
wheels: they were among
the fastest locomotives of
the age, with top speeds of
up to 60 miles per hour. The
Argus *was in service until*
1873.
[Great Western Trust
Collection]

locomotives: they were underpowered and unreliable, and most historians have concluded that Brunel bore the main responsibility for this.[56] As a result, when the railway opened, the only really reliable locomotive it had was the *North Star*, built by Robert Stephenson & Co in Newcastle upon Tyne for the 5ft 6in. (1.68m)-gauge New Orleans Railway and converted to the 7ft (2.14m) gauge for the GWR after the American company went bankrupt.[57]

The GWR's mechanical reputation (and probably its business) were saved by Daniel Gooch, whom Brunel appointed as the company's locomotive superintendent at the age of 21 in August 1837, before the company actually had any engines to superintend. Gooch got Brunel's collection of freak locomotives running but with great difficulty. As Charles Saunders, the company secretary, put it: 'At the opening in 1838 we found the engines were so inefficient that time-table working was hopeless; one or two engines might keep time, the other eight or ten were always out of time. So we suspended time-tables till the locomotive power became sufficient.'[58]

These problems caused great tensions between the board, Gooch and Brunel, who was anxious to maintain his primacy in the company.[59] By 1839 Gooch was effectively given his independence and was responsible for the second generation of GWR locomotives,

commissioning a series based on the *North Star* and another series, the *Firefly* class, from Stephenson & Co and other makers (Fig 1.10).[60] In September 1840 Gooch proposed Swindon in Wiltshire as the site for the company's main engine depot and workshops. New buildings, including a 'railway village', went up there in 1841–3 and in 1845 the works were expanded to allow the GWR to build its own locomotives. The first engine was completed in February 1846, the start of 140 years of locomotive building at Swindon.[61] Here Gooch masterminded and produced the fastest and most powerful locomotives of the age, the '7ft singles' and the celebrated '8ft singles', named from the diameter of their great driving wheels. In its first years, the GWR's trains were not especially fast, averaging 33 miles (53km) an hour, rather better than the London & Birmingham's 27 miles (43km) an hour. Official criticism of the broad gauge in 1845–6 stung them into action. The company laid on express services travelling from London to Bristol in four and a half hours: between Paddington and Swindon their average speed was 59 miles (95km) an hour, twice that of any other railway's services. The combination of Brunel's track and Gooch's engines gave the Great Western Railway almost the fastest and probably the safest trains in the world.[62]

The First London Terminus, 1835–1850

Paddington was one of the first sites that the Great Western Railway considered for their all-important London terminus (Fig 2.1). When Brunel presented his survey to the first public meeting of the London & Bristol Railway (as it was still called) on 30 July 1833, his report said: 'The total length of the Railway would be from 115 to 118 or 120 miles [185, 190 or 193km], depending on its termination whether at Paddington, or on some part of the Southern bank of the Thames.'[1]

Finding a viable route into London and persuading all the relevant landowners to sell was clearly going to be difficult. The prospectus issued at the time of the first board meeting in August 1833 showed two alternatives, one terminating in Paddington, and the other crossing the Thames below Kingston and curving up the south bank, ending between Westminster and Waterloo Bridges.[2] The plans deposited with the first parliamentary bill in November, however, showed a third alternative: a line from South Acton through Hammersmith and Pimlico, ending at Vauxhall Bridge, between Vauxhall Bridge Road and the Kensington Canal.[3] The company engaged in inconclusive negotiations with Lord Cadogan, owner of much of the land required. The bill, as we have seen, was defeated in the House of Lords on 25 July 1834.[4]

The GWR went back to the drawing board. On 2 September 1834 the directors discussed the question of a London terminus with Brunel and decided to approach the London & Birmingham Railway 'for the purpose of ascertaining and determining whether a junction

*Figure 2.1
'Interior of the Great Western Railway Terminus Paddington', a print of 1843 showing the covered roadway between the arrival and departure platforms.
[City of Westminster Archives Centre]*

might be made with their line at or near Harlesden Green'.[5] The next edition of the GWR's prospectus, in November 1834, said: 'The line of railway … will be 114 miles [183km] in length from Bristol to the point of junction with the Birmingham Line near Wormwood Scrubs. The Station for Passengers in London is intended to be near the New Road in the Parish of St Pancras.'[6]

In other words, a site just north of Euston Square, named for the Suffolk country seat of the Duke of Grafton, who owned much of the area. In this period, before the Great Western and the London & Birmingham were in direct competition with each other, the idea of sharing the enormous cost of a London terminus doubtless had major attractions. The London & Birmingham were ahead of the GWR: their original Act of 1833 specified a terminus at Chalk Farm, but they secured more land running south to the New Road (now the Euston Road) and obtained an Act of Parliament for an extension in 1835.[7] The GWR's directors were certainly serious about the idea. It was written into their new parliamentary bill, and when the Great Western Railway Act received royal assent in August 1835 it named Euston as their intended terminus.

Strangely, given the vital importance of the issue, the GWR and the London & Birmingham Railway do not seem to have started discussing the joint station in any detail until October 1835. From then until March 1836 they engaged in tortuous negotiations, ultimately fruitless, with something of the character of a poker game.[8] The GWR, understandably, wanted to own their half of Euston Station outright. This met with a point-blank refusal from the London & Birmingham board. The GWR then asked for a 21-year lease, the bare minimum they regarded as acceptable to protect their interests.

Around the beginning of November 1835 Brunel and Robert Stephenson, the London & Birmingham's engineer, met to produce outline proposals for the joint station. On 19 November, Brunel explained these to the GWR directors. The Euston site was fairly narrow from east to west and could only accommodate the passenger terminus: everything else would have to be further north at Chalk Farm and Camden Town. The land at Euston Square would be divided longitudinally: the western half would be the Great Western's, the eastern half the London & Birmingham's. The 'façade or public front of the joint depot' fronting Euston Square would be designed by a 'joint architect appointed by both companies'. The site sloped downhill from Chalk Farm, south towards Euston Square, and this slope was too steep for the primitive locomotives then available. Stationary steam engines would be needed at Chalk Farm to tow the passenger trains up the inclined plane. The companies would have to share the inclined plane and the stationary engines, but the London & Birmingham would construct and maintain them. Land would be set aside at Camden Town for the GWR's engine sheds, goods depot and warehouse. Cattle yards and sidings, though, would be held in common and shared. Brunel concluded by expressing scepticism about the wisdom of any leasehold arrangements.[9]

It became clear, however, that 'a large and influential body of distant proprietors' – in other words the 'Liverpool people' or 'Liverpool party', the London & Birmingham's powerful northern shareholders – were lobbying for a shorter lease, perhaps as short as five years. The Liverpool party were buying GWR shares on a large scale and the Great Western's directors were becoming very wary of them. They were not sure whether the short lease was intended to give the London & Birmingham Railway control over the GWR, or whether it arose from concerns that the two companies would outgrow Euston, and allow the L&BR to eject the GWR in short order.[10]

The gauge issue also stood between the two companies. Brunel seems to have been pondering the question from around the end of 1834. Surprisingly, he did not share his thoughts with the GWR directors until 15 September 1835, when he wrote them a long letter on the subject.[11] Brunel carried the board with him and on 29 October they agreed to the broad gauge. Yet at the very same board meeting 'Mr Brunel reported that he had had an interview with Mr Robt. Stephenson the Engineer of the London and Birmingham Railway Company, who expressed his impression that he would advise the Directors of the London and Birmingham Railway Company not to accede to the proposals to lay rails of a greater span on that railway.'[12]

On 21 November the GWR's London Committee heard unofficially about a private meeting of the Liverpool party. What they heard alarmed them so much that on the same day they ordered Brunel to proceed with surveys for a different route into London, to

Paddington. Another meeting with the London & Birmingham's secretary on 26 November seemed to put the negotiations back on track: the GWR's London Committee advised that the Paddington route could be dropped again.[13] The day before, the London & Birmingham had let a contract to William and Lewis Cubitt for building their last section of line to Euston, which may have had something to do with this softening in their position. Through December and January into February, discussions circled around the capacity of the sites at Camden and Euston, the rates that the GWR would be charged for use of the line, designs for the carriages they would operate jointly and so on. But the London & Birmingham would still not finalise the terms of the lease.

Thus it was that the GWR secretly revived the Paddington idea. On 9 February 1836 Brunel reported that arrangements to acquire land there had been made: they would have to proceed with 'great precaution', as any money spent on the Paddington extension would be at risk until they got the necessary Act of Parliament.[14] On 10 February 1836 C A Saunders, the London Committee's secretary, wrote to the London & Birmingham board, repeating the GWR's request for a 21-year lease. The board met on 17 February: they offered five years. For the GWR this was completely unacceptable. A decision could not be put off much longer, as a half-yearly shareholders' meeting was already set for 26 February 1836. The GWR directors told the meeting that the London & Birmingham Railway would not concede the 21-year lease, but they believed that 'an excellent and

independent terminus can be secured without much difficulty at latest in the next session of Parliament'.[15]

The Liverpool party, who were well represented amongst the GWR's shareholders, would have been in the audience. The negotiations collapsed. Saunders' letter to the London & Birmingham directors, rejecting the five-year offer, was read out at their meeting on 2 March.

Nevertheless, the London & Birmingham continued with their plans for Euston, almost as if nothing had happened. In 1836 their architect Philip Hardwick produced designs for a double station in line with Brunel and Stephenson's ideas. This was the origin of Euston's rather curious plan. The famous portico (or more accurately propylaeum) known as Euston Arch was to be aligned with a central roadway running northwards for the length of the station (Fig 2.2). Twin passenger stations would be arranged symmetrically on either side, the GWR notionally occupying the left (west) half, the London & Birmingham the right.[16] Hardwick's design for Euston has been criticised on the grounds that the magnificent arch and façade bore no relation to the buildings and functions behind, but this is unjust.[17] His design makes perfect sense if understood as a grand entrance aligned with a central axis, with twin stations on either side.

It seems strange that the London & Birmingham went ahead and built Euston to this plan, well after the negotiations had broken down. Their London Committee considered the issue on 14 September 1836. Stephenson advised them that one half of the station, as

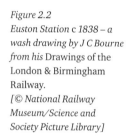

Figure 2.2
Euston Station c 1838 – a wash drawing by J C Bourne from his Drawings of the London & Birmingham Railway.
[© National Railway Museum/Science and Society Picture Library]

originally planned, would suffice for all their present needs, so they resolved that 'the plan of a double passenger station of four lines to each wing, but of which only the East wing is to be constructed for the present, be approved and adopted' (Fig 2.3).[18]

On 5 December 1836 Hardwick went north to secure the approval of the Liverpool shareholders.[19] The contract for the Euston arch, lodges and gates was let to William and Lewis Cubitt on 7 December, and the contract for the arrival and departure stages, drains, boundary walls and paving on 5 April 1837.[20] The plans of Euston as it opened in June 1838 look bizarre, with the western half of the station left empty. It represented growth room, but it is also possible that the London & Birmingham Railway built Euston to this plan with a view to taking the Great Western Railway over, and bringing their line into Euston after all. The cost of building the GWR was rising far beyond the initial estimates, the company had to raise more share capital, and in 1836–8 the Liverpool party were buying into it on a large scale. Through the second half of 1838 they used their voting power in the GWR to lobby against Brunel. In the crucial vote on 7 January 1839 over whether to continue with the broad gauge, Brunel and Saunders' opponents were only defeated by 7,792 votes to 6,145.[21] Had the Liverpool party been victorious, Brunel and Saunders would have been dismissed, the Great Western would have been converted to the standard gauge, and

its line might have been brought into Euston after all, as provided for in the GWR's original Act of Parliament. Paddington Station might have closed almost as soon as it had opened.[22]

As it turned out, the London & Birmingham Railway filled Euston within 10 years, and it would certainly not have sufficed for both companies. In 1845–6 the London & North Western Railway, as they had just become, filled the empty half of their station with P C Hardwick's great hall, boardroom and shareholders' meeting room, creating an asymmetry in Euston's layout which persisted until its demolition in 1961.

The GWR had a problem. Their Act of Parliament authorised them to build a line to Euston but they no longer wanted to do this, or at any rate, they had been unable to negotiate satisfactory terms. Work had already started on their line in November 1835 at Hanwell to the west of Ealing, where the resident engineer J W Hammond and the contractors Grissell & Peto began work on the foundations of the great viaduct over the Brent valley later known as the Wharncliffe Viaduct.[23] A mile or so to the east of Ealing, between Wormwood Scrubs and Kensal Green, there is a point where today the Great Western main line and the West Coast (originally the London & Birmingham) main line are less than a quarter-mile apart: under the GWR's original Act, their line was supposed to swing north at this point, and follow the London & Birmingham's line into Euston.

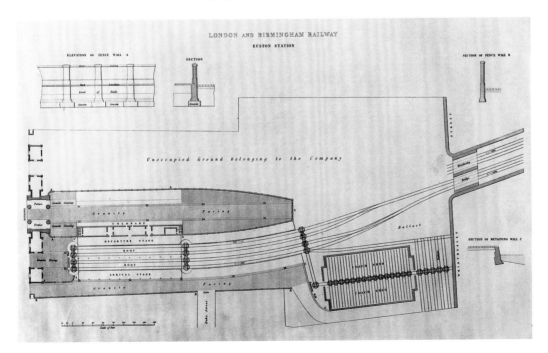

Figure 2.3
Philip Hardwick's original plan for Euston Station, showing its intended dual layout. The land to the west of the central axis was originally intended for the Great Western Railway.
[© The British Library Board (650.c.17/74)]

Taking their line to Paddington instead would require a new Act of Parliament and the 1836 session of Parliament was too far advanced for them to start a bill that year. In the event, their Extension Act was not passed until July 1837.

The site Brunel had selected was next to the Paddington basin of the Grand Junction Canal which had opened in 1801: the prospect of establishing a direct link to the canal was doubtless an important factor in his choice (Fig 2.4).[24] Most of the immediate area belonged to the Bishop of London's estate. In 1836 London's western suburbs were spreading along the north side of Hyde Park and up the Harrow Road and in about 1828 the bishop's surveyors, Samuel Pepys Cockerell and his successor George Gutch had begun to lay out streets and building plots in the area.[25] The 'enormously high price' that the GWR had to pay the bishop's trustees doubtless reflected the fact that the area was on the verge of being developed.[26] The GWR had to buy plots from several other owners, including the Grand Junction Canal Company, Paddington Parish Vestry and John Cockerell, owner of Westbourne Place, a large villa in extensive grounds which at the time was let to Lord Hill, who had been one of the Duke of Wellington's principal commanders.[27]

The Paddington Parish Vestry seem to have regarded the railway as a good thing, but they

and the other vendors must have known that they were in a strong bargaining position and the negotiations took a long time.[28] The company only needed to buy a small plot of land from the vestry worth £1,200, but they also needed the vestry's consent to their scheme and this involved fulfilling several conditions. An outline agreement was reached with the vestry on 4 February 1837,[29] and with the bishop's trustees on 16 February 1837.[30] The largest plot was an area of just over 44 acres (17.8ha) leased from the bishop's estate for an initial payment of £30,000 and an annual rent of £2,366 12s. The Great Western Railway never owned the freehold of the site of Paddington: the Church Commissioners, to whom the estates had passed, eventually sold it to the British Transport Commission in 1958.[31] The second major plot was bought from the Canal Company for £24,109. In 1838 the GWR estimated that in total they had spent £175,000 on purchasing property and compensating owners for the extension to Paddington, including the line, depot and station.[32]

The site at Paddington was in a shallow depression or bowl. This was probably the main reason that Brunel chose it in the first place, to allow for the easiest possible gradient, and allow the railway to pass below street level. The Paddington Parish Vestry generally welcomed

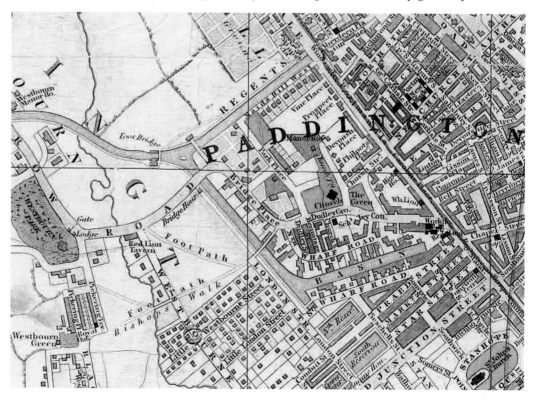

Figure 2.4
The Paddington area c 1838, from Crutchley's map of London. The site of the station extends from Bishop's Walk south-east to Conduit Street, to the left of the canal.
[© The British Library Board (Maps Sheet 1 4379 (43))]

the railway, but insisted that a number of brick viaducts be built to prevent their district from being cut in two.[33] The longest of these, linking Spring Street (now Eastbourne Terrace) to the Harrow Road, became known as the Bishop's Road Bridge. A little to the west a new viaduct, the Westbourne Bridge, was needed to carry the newly laid-out Westbourne Terrace north to join the Harrow Road. West of this another new viaduct, the Ranelagh Bridge, would link the Gloucester Road and Porchester Road to the Harrow Road. Finally, a fourth viaduct was needed to carry Black Lion Lane: this was styled Lord Hill's Bridge after the area's most distinguished resident. These four viaducts represented a large part of the total cost of the Paddington extension. In the event the Bishop's Road Bridge was used as the framework for the first 'temporary' station at Paddington.

This was certainly not Brunel's original intention. One of his surviving sketchbooks, dated 1836, contains a number of sketch plans and elevations labelled 'GWR Depot' for a terminus at Paddington (Fig 2.5).[34] They provide a partial picture of what Brunel intended and they represent a lost chapter in the history of railway station design. The general plan (Fig 2.6) shows the passenger terminus sitting on the present site and entered from Conduit (now Praed) Street, with a goods depot, on the site that it was eventually to occupy, planned around a new basin opening off the Grand Junction Canal. Where the passenger station is concerned, the sketchbooks show minor variations but all three versions agree on the main outlines. A broad central roadway marked 'private carriages' ran down the middle of the station with, in one version, stables alongside it. It runs into a turning space with rounded corners: a finished drawing shows several tracks running up to the platform edge here. We can interpret this as being a loading dock, where the horses and carriages of wealthy passengers could be loaded onto horseboxes and carriage trucks. To either side of this central roadway are broad trackbeds and on the outer sides there seem to be platforms. On the south side, one of the sketches has areas labelled '1st Class' and '2nd Class', which may be taken to represent the main 'departure platform'. The 'arrival platform', presumably, would have been on the north side.

Brunel sketched out two alternative designs for the main façade to Conduit Street. There would be a grand entrance to the road for private carriages, one version with a big central

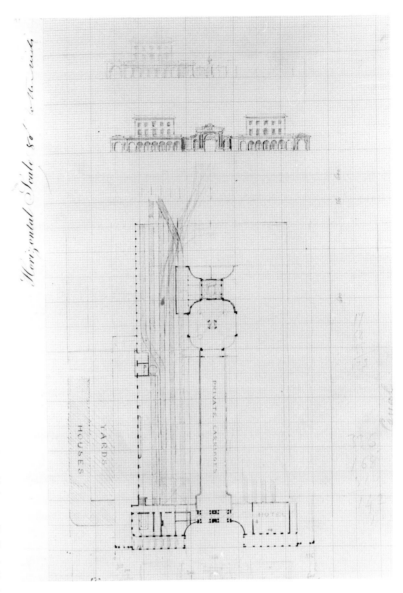

arch flanked by smaller arches, another version with a pair of identical arches. The arches would have been flanked by quadrant walls, arcades and five-bay buildings, all in an Italianate Classical style. On these drawings Brunel quite carefully marked the tracks running right up to the rear wall of the buildings. The implication seems to be that Brunel was planning to exploit the difference in level between the street and the tracks, to run the tracks under Conduit Street and towards Central London in the future. This chimes in with a reference in 1845 when Brunel, in discussing the site of the goods depot, recommended leaving room 'in order to carry lines of railway farther into London hereafter'.[35]

When Brunel produced this first design, only two proper railway termini had been built (in

Figure 2.5
Brunel's first sketch for a station at Paddington, December 1836. A grand archway leads to a central roadway with platforms on either side, rather like the intended dual terminus at Euston.
[University of Bristol Brunel Collection (Sketchbook GWR 1836, 47). By courtesy of the Brunel Institute, a collaboration of the SS Great Britain Trust and the University of Bristol]

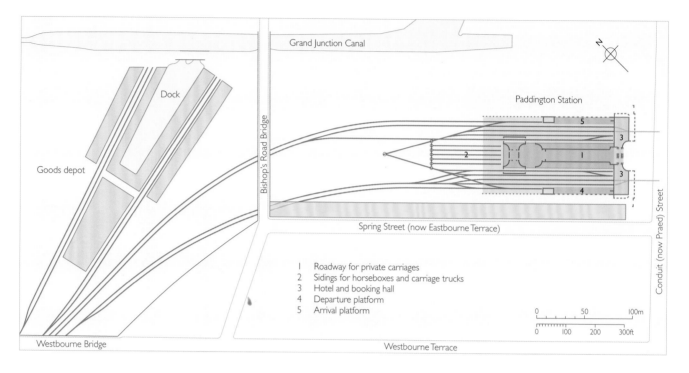

Grand Junction Canal

Dock

Paddington Station

5

3

1

2

3

Goods depot

4

Bishop's Road Bridge

Conduit (now Praed) Street

Spring Street (now Eastbourne Terrace)

1 Roadway for private carriages
2 Sidings for horseboxes and carriage trucks
3 Hotel and booking hall
4 Departure platform
5 Arrival platform

0   50   100m
0   100   200   300ft

Westbourne Bridge

Westbourne Terrace

Figure 2.6
Brunel's first master plan for Paddington, December 1836, redrawn from a series of contemporary plans.
[© English Heritage]

Liverpool and Manchester): the building type was only just being developed. This scheme, with its grand entrance and tracks disposed to either side of a central roadway, bears a marked resemblance to Philip Hardwick's original plan for Euston, which of course had developed out of Brunel and Robert Stephenson's negotiations. This seems unlikely to be a coincidence.

Another striking point about the design is the prodigally generous provision for the owners of those private carriages, who seem to have been the sole beneficiaries of the grand entrance: it looks as if all the other passengers, whether first or second class, would have been directed to the more modestly scaled outer platforms. At this stage, Brunel seems to have been envisaging that the private carriages would be loaded onto trucks, and their wealthy passengers would travel in their own carriages, as indeed happened when the GWR opened. Brunel's design seems to be a vivid expression of the GWR's sense of social decorum, and for whom it was really built: the grand entrance was for persons of quality, arriving and probably travelling in their own carriages, while everyone else could go round the side. In late Georgian Britain, after all, perhaps the most visible and important social marker was whether you kept your own carriage.

By July 1837, when the GWR obtained the Act of Parliament for their extension to Paddington, it had become clear that their main line was going to cost vastly more than the £3.3 million authorised by their original Act. Brunel was driven to economise in all directions and his plan for a grand London terminus was among the casualties.

Work on the line from London to Maidenhead was going well and the directors were keen (perhaps rather too keen) to push the work forward. At a half-yearly shareholders' meeting on 31 August 1837 they announced that the line would open as far as Maidenhead in November of that year. As for the terminus at Paddington, they reported that 'the space allotted for the Depot in London, for Goods, will for the present be employed as a Temporary Station for Passengers until the permanent buildings can be erected'.[36]

The company had already started excavating the site in April 1837. Michael Tutton has reconstructed the sequence of events from the GWR's account books.[37] Until September 1838, all the relevant payments come under the heading 'Paddington Terminus'. As Tutton notes, this seems initially to mean the embankments, retaining walls, bridges, drains and roadworks. The GWR used their directly employed workforce for the excavation works, spending £36,870 on wages in 1837 and another £33,903 in 1838–9. The brickwork of bridges, retaining walls and the station itself was divided up into contracts, which were let to two of the largest and most reliable contractors, Grissell &

Peto (later to build the Palace of Westminster) and B & N Sherwood.[38] The two firms were paid £30,030 in 1837 and another £36,381 in 1838–9.[39] Considerable sums were also being spent on timber, rails, wheels, wagons, stone and gravel.

Brunel drove the work forward at a frantic pace, but the board's November 1837 deadline was completely unrealistic. The GWR only took delivery of their first locomotives in November, and at that point there was still not a single mile of track for them to run on. By Christmas, 1½ miles (2.4km) of track had been laid between Drayton and Langley. By 12 April 1838, there was still only 5½ miles (8.8km). Work must have speeded up greatly, for the line was finished from Paddington to Maidenhead, a distance of 22 miles and 43 chains (36.3km), by 31 May. On that memorable day the directors met at Paddington at 11 am and travelled to Maidenhead in 49 minutes in a train pulled by the *North Star* at an average speed of 28 miles (45km) an hour. They then inspected work on the bridge over the Thames there, had a cold luncheon at Salt Hill a couple of miles back down the line near Slough, returning to Paddington at a brisk 33½ miles (54km) an hour. The Great Western Railway opened to the public at 8 am on Monday 4 June 1838.[40] The company did not really have a station at Paddington, which was still a building site and would remain so into the 1840s, and it is not clear how the first services and the first passengers were accommodated.

On 15 August 1838, with the first stage of their line open, the board issued estimates for the cost of their extension from Acton to Paddington. They had spent £175,000 on land and compensation: 39 acres (16ha) for the station and its outworks, and 156 acres (63ha) for the 4½ miles (7.2km) of line. They had excavated 580,000cu yd (443,000cu m) of earth for the track and its related embankments and roadworks, at a cost of £56,444. They expected to spend £62,508 on brickwork, including culverts and sewers. Another £83,366 came under the general heading of 'Paddington extension works'. The permanent way, which was already finished and open, had cost £41,178 (at £9,173 per mile). For buildings, including the passenger sheds, coach house, engine house and all the mechanical fittings, they planned to spend a relatively small £27,328. They expected the extension into London to cost a total of £450,224.[41] The accounts recorded a comparatively modest £5,439 spent on 'Paddington Station' between September 1838 and December 1839, referring to general construction costs, fitting out, ironwork, lamp posts, laying pipes and machinery. Work began on the engine house in September 1839 (it cost £1,602) and on the merchandise station or goods shed in October 1839 (which cost £3,802).[42]

By 1845 a fully functioning station was in place (Fig 2.7). Roads were made leading down from Conduit Street and London Street to the station approach, and simple but dignified gates built at the entrance: a sketch in one of Brunel's sketchbooks shows that, characteristically, he designed these himself.[43] They led into a large, irregularly shaped yard on the site occupied by the present passenger station. Brunel had, indeed, planned to put his passenger station there, until the need for economy ruled this out. Instead, he put the temporary goods depot there: this was a wide, relatively lightly constructed shed, 330ft (100m) long by 120ft (37m) wide; early views show it having timber-framed and boarded sides (Fig 2.8).[44] The goods offices were off to the left, against the slope: they looked rather like a converted cottage (*see* Fig 2.13).

Directly ahead, spanning the low valley or cutting, was the Bishop's Road Bridge, which formed the station's principal façade (Fig 2.9). There were no fewer than 26 arches, with spans varying from 17ft (5.2m) to 20ft (6m), providing a fair amount of accommodation. The first nine arches were all skewed: reading from the left (south) side, tracks ran through the first three arches to the goods shed and a head-shunt; the next three arches also housed goods sidings; of the final three, two seem to have been blocked and the third was left open for carts taking luggage through to the departure platforms.

Beyond this point, the remaining arches were at a conventional right angle to the façade. The first nine of them formed a symmetrical design: this is the part which Bourne concentrated on in his lithograph and which housed the station entrance. The first arch, with a veranda in front of it, housed the booking hall. The next two were waiting rooms, the fourth one the 'down' parcels office. The fifth and central arch of the nine was the carriage entrance, through which the GWR's richer passengers could drive straight onto the departure platform, to have their horses and carriages loaded onto a train. The sixth arch was the carriage exit, the next

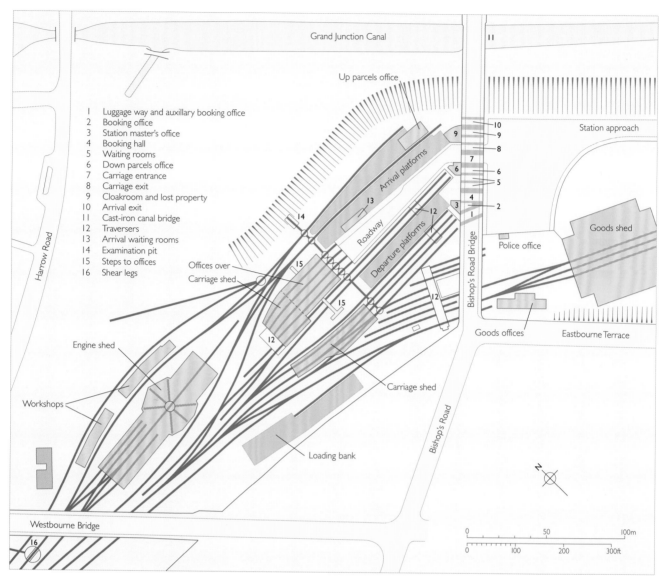

1   Luggage way and auxillary booking office
2   Booking office
3   Station master's office
4   Booking hall
5   Waiting rooms
6   Down parcels office
7   Carriage entrance
8   Carriage exit
9   Cloakroom and lost property
10 Arrival exit
11 Cast-iron canal bridge
12 Traversers
13 Arrival waiting rooms
14 Examination pit
15 Steps to offices
16 Shear legs

Figure 2.7
Paddington Station in 1845, redrawn from a plan accompanying a series of historical articles in the GWR Magazine (Matthews 1916).
[© English Heritage]

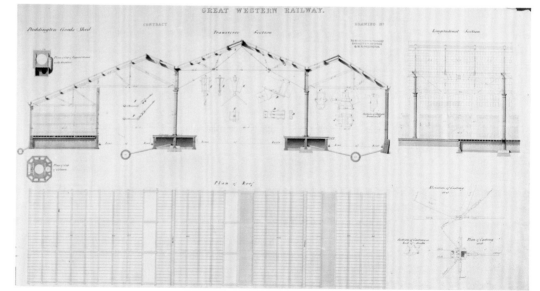

Figure 2.8
A design for the first goods shed, c 1838, cross section.
[Network Rail/EH AA031050]

Figure 2.9
Paddington Station c *1846,*
by J C Bourne from his
History and Description of
the Great Western Railway.
[STEAM – Museum of the
Great Western Railway,
Swindon]

two were cloakrooms, and the last of this symmetrical group of nine was the entrance to the arrival platform. Further to the north, six more brick arches took the bridge up to the canal, and one more iron span crossed it (*see* pp 151–7).

On the other, western, side of the viaduct was a complex network of lines, with timber platforms covered by simple roofs carried on iron columns (Fig 2.10; *see also* Fig 2.1). It must have been rather confusing for the visitor and can

never have looked very designed. The southern departure side had three lines of track served by two fairly narrow platforms.[45] Next there was a broader roadway for carriages, ending in a platform edge and a series of wagon turntables. These allowed wagons to be run close up to the platform edge, so that horses could be led onto horseboxes and carriages wheeled onto carriage trucks: this was an important part of the GWR's business (*see* chapter 5). Further

Figure 2.10
Inside the first Paddington
Station: a watercolour of
c *1840. This seems to show*
the northern or arrival side
looking south-west.
[© National Railway
Museum/Science and
Society Picture Library]

north still were platforms flanking the two arrival lines.[46]

A little to the west, just beyond the horse and carriage dock, was the main carriage shed. Measuring 180ft (55m) by 70ft (21m) and spanning four lines of track, it was added in July 1840 at a further cost of £2,446.[47] Off to the left and south was another carriage shed. On the southern edge of the site, lines ran through to the first three arches and the goods depot, as described earlier. There was also a long traverser, linking the goods sidings together. Traversers, or traversing frames, which Brunel had apparently developed during the construction of the GWR in c 1837–8, became standard pieces of equipment. They comprised mobile sections of track, carried on wheeled iron frames and running on rails at 90 degrees to the fixed track, which allowed carriages or wagons to be moved sideways from one line to another. Traversers were certainly installed at Reading in 1839 and at Bristol Temple Meads between c 1841–4, and they subsequently appeared in many other GWR stations.[48] The early ones all seem to have been manually operated with capstans: they were meant for carrying carriages and trucks, which were still quite light.[49]

Further to the west towards Lord Hill's Bridge was the engine house, the main part of which was a polygonal, iron-framed shed (Fig 2.11). This was designed by Daniel Gooch, the young locomotive superintendent who had joined the company in August 1837: in his memoirs he wrote that, not yet having any locomotives to look after, his first task had been to design the engine houses at Paddington and Maidenhead.[50] Gooch therefore seems to have a claim to have invented the roundhouse design for engine sheds, the next example being Francis Thompson's polygonal shed at Derby for the North Midland Railway of 1840.[51] Gooch's engine house at Paddington was demolished in the 1850s, and is known only from plans and one engraving.

Work on the passenger station, engine house, goods station and stables continued through 1840–1. Everything had been done in a hasty and rather haphazard way. It is not surprising that several improvements were soon needed. The engine house was extended in 1840–1 (one of Brunel's volumes of 'Facts' has an interesting sketch design for load-testing its roof trusses).[52] On 19 March 1842 the GWR board decided to improve the platform facilities and add a waiting room, and on 8 April they decided to roof over the whole space between the arrival and departure platforms, for which an estimate of £475 was duly approved.

Meanwhile the GWR had been reforming the way it ran itself. In 1841 the London and Bristol Committees which had overseen the construction of the main line were abolished and replaced with the Traffic and General Committees. The board itself was reduced in number from 24 to 18 (February 1842). The company briefly flirted with the idea of making Steventon in Berkshire its head office on the grounds that it was midway between London and Bristol, and a boardroom was fitted up in the superintendent's house there. This did not last long and early in 1843 they decided to move their headquarters to Paddington.[53] On

Figure 2.11
The engine house designed by Daniel Gooch, probably the first to be built on the roundhouse plan.
[Great Western Trust Collection]

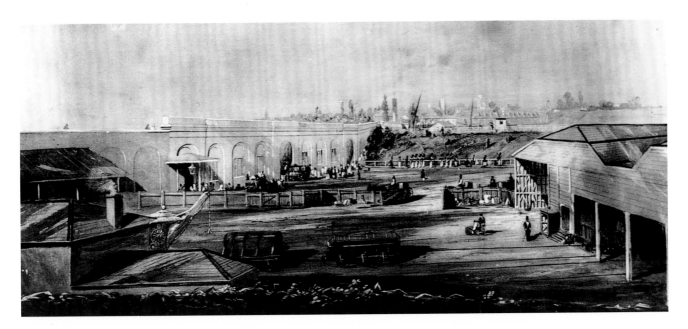

4 March Brunel submitted plans for a building to go over the main carriage shed to house additional offices, at an estimated cost of £2,400, and work began on this in August.

Thus the first Paddington Station was a makeshift affair (Figs 2.12 and 2.13). Modestly sunk in its cutting, it could not compete architecturally with the London & Birmingham Railway's magnificent façade at Euston. J C Bourne, drawing it for his 1846 *History and Description of the Great Western Railway* (*see* Fig 2.9), managed to give it a certain simple dignity by concentrating on the symmetrical part of the Bishop's Road Bridge and by some judicious omissions. Other early views give a somewhat fuller picture. Only two views of the interior (to the west of the bridge) are known, showing the simple iron columns and roofs.[54]

Paddington Parish Vestry had initially given the GWR a guarded welcome, but like other

Figure 2.12
The first terminus at Paddington – an oil painting of c 1840.
[City of Westminster Archives Centre]

Figure 2.13
Paddington Station in the 1840s, a view looking south towards the Prince of Wales' Hotel, with the cottage-like goods office in the foreground, emphasising the informal quality of the first terminus.
[STEAM – Museum of the Great Western Railway, Swindon]

communities, they had probably reckoned without the problems that started to arise, with this new centre of activity suddenly planted in the middle of their neighbourhood:

Many and heavy complaints have been made to the Vestry of the Nuisances which are daily and hourly committed on the Bridge over the Canal, and at the corner of North Wharf Rd, principally by Passengers to and from the Station, owing, as the Vestry conceive, to the entire want of proper conveniences on the Premises of the GWRC, and I am to suggest the necessity of such conveniences being immediately erected, to remedy the evil complained of, the Boards recently stationed on the Bridge by the Vestry proving ineffectual … .

They also complained of 'the seeming unnecessary and protracted duration of the whistles of the engines, which it is hoped can be suppressed'.[55]

When the GWR opened their line to Maidenhead in June 1838, they can have had little idea of the level of demand. In the first week 10,360 people passed through the makeshift terminus and the company took £1,552 (*see* Table 1).[56]

Table 1 Fares when the line to Maidenhead opened

	First Class		Second Class	
	Posting Carriage	Passengers Coach	Coach	Open Carriage
West Drayton	4s 0d	3s 6d	2s 0d	1s 6d
Slough	5s 6d	4s 6d	3s 0d	2s 6d
Maidenhead	6s 6d	5s 6d	4s 0d	3s 6d

The company started running goods trains in 1839, operated by carriers, who also took third-class passengers in the goods wagons. From 1840, the GWR began offering this rather uncomfortable service themselves: 'the Goods Train Passengers will be conveyed in uncovered trucks by the Goods Trains only'.[57]

By 1840, the company had enough of Gooch's reliable new engines to introduce proper timetables and also to start running a night mail service. They had also made their first experiment in providing 'special trains' when they laid on a locomotive with a single carriage to take the Duke of Lucca to Windsor in January 1839, at a cost of £10.[58] They carried the new Prince Consort between Windsor and London a number of times, and in 1840, anticipating further royal patronage, they commissioned a royal carriage 'most magnificently fitted by Mr Webb, the upholsterer, of Old Bond

Street'. On 13 June 1842 Queen Victoria made her first railway journey, from Slough to Paddington. It seems to have been a success and before long the queen was a regular passenger on the GWR.[59]

In 1843 the company surveyed their operations with a view to cutting their costs. They counted 213 employees at Paddington, plus 4 conductors, 42 passenger guards and 10 goods guards based there – a total of 269. Among these was Mary Coulsell, the first woman to be employed by the company, appointed as the 'female attendant at Paddington' in May 1838.[60]

One important development on the GWR in these early years was the installation of the first electric telegraph line to be laid and used over any major distance, in 1838–9. The invention had only been patented in 1837. Its inventors, William Fothergill Cooke and Charles Wheatstone, had made experiments on the Liverpool & Manchester Railway, but that company had already committed itself to a pneumatic tube communication system. They offered their invention to the London & Birmingham but their board would not accept it. Brunel was enthusiastic and persuaded the GWR board to lay telegraph lines for the 13 miles (21km) from Paddington to West Drayton. The line cost £2,817 10s and was an immediate success, carrying general messages and information about the running of trains.[61] The telegraph attracted immediate interest, and on 24 August 1839 Seymour Clarke, the traffic superintendent, wrote to C A Saunders, the company secretary: 'The Duke of Wellington, Lord Bathurst, Lord Fitzroy Somerset, and some ladies and afterwards Lord Howick came. The telegraph worked capitally. Mr Sims and Mr Gibbs [GWR directors] were here. The Duke appeared much pleased.'[62]

In 1845 the GWR's telegraph system achieved still greater fame, by being instrumental in helping the police to catch a murderer. Cooke and Wheatstone sold their patents to the Electric Telegraph Company, which ran the GWR's system to Slough in 1843, but removed it after a dispute in 1849. In 1852 the telegraph was reinstalled over the whole distance from Paddington to Bristol.[63]

The GWR's network and traffic grew apace in the 1840s and the board soon found themselves faced with the inadequacy of their temporary station. On 6 November 1845 they discussed the building of a 'permanent goods station' at the end 'nearest Conduit Street' (towards the site of the present hotel). Brunel supported the

Figure 2.14
The earliest known
photograph of Paddington,
half of a daguerrotype
'stereo pair'. It shows the
Westbourne Bridge from the
south, (the shear legs are
shown on Fig 2.7 at number
16). The roof of the first
engine shed is visible beyond
the bridge. Demolition of the
first station seems to be in
progress, hence the salvaged
materials lying around,
dating it to c 1854–5.
[Reproduced by permission
of English Heritage
Howarth-Loomes Collection
BB72/450B]

idea but advised that space be left 'in order to carry lines of railway farther into London hereafter in case such measures be adopted'.[64] He was instructed to prepare plans, but nothing seems to have been done.[65] The board returned to the subject in July 1846 and asked Brunel for plans. On 3 August he presented two options. One was to 'widen and lengthen' the existing station and replace the 'merchandize station' on land adjoining Spring Street (now Eastbourne Terrace, so on the site of the present station). The other plan was to build a new passenger station on the Eastbourne Terrace site and convert the temporary station for goods. Again, neither proposal was taken up. An undated plan in Network Rail's collection shows a different scheme for improving the station by adding an entrance building alongside the bridge, to house a round entrance hall, waiting rooms and a booking hall at bridge level, with stairs down to the platforms.[66] It would not have increased the station's real capacity and this too was never carried out.

Brunel's first station was demolished in *c* 1853 (Fig 2.14), and the southern section of the Bishop's Road Bridge was demolished and replaced with a hog-back truss in 1906. The five northern arches of the bridge stood until 2004, when the whole structure was demolished to make way for a new four-lane road bridge. The northernmost section of the Bishop's Road Bridge, spanning the canal, was recognised as Brunel's earliest surviving iron bridge and salvaged at this time (*see* chapter 8).[67]

3

Building the New Station, 1850–1855

The Great Western Railway's network and traffic grew steadily through the 1840s and so did the pressure on the temporary station at the Bishop's Road Bridge: the company was obliged to make additions and alterations to it every year. In 1847 it was new carriage shops and sheds, in 1848 a large new engine house, and in 1849–50 it was new grain stores, the re-laying of a timber platform, roof repairs and repairs to the engine house and stables.[1]

By 1848 work was under way on the Oxford & Rugby Railway, the Birmingham & Oxford Junction Railway and the Birmingham, Wolverhampton & Dudley Railway. These GWR satellite companies, between them, would take the broad gauge to Birmingham and beyond, putting the Great Western in direct competition with the London & North Western Railway. This promised a huge further increase in the volume of traffic to Paddington. Yet still the company moved very cautiously. On 18 April 1850 the board of directors wrote in their six-monthly report to shareholders that the station was indeed inadequate, but 'to provide such accommodation before the immediate necessity arises would, as the Directors think, be imprudent, and they merely refer to it now, to guard themselves against cavils or reproof'.[2]

The tone of extreme caution and the reluctance to make a major decision can readily be explained. The whole railway industry had gone into financial crisis in the late 1840s. The boom in railway shares orchestrated by the 'railway king' George Hudson had come to a dramatic end, all the main companies' shares fell in 1847–9, and in 1849 the GWR was obliged to tell an angry shareholders' meeting that it was reducing its dividend from 4 per cent to 2 per cent.[3] The board would have been aware of the lavish new shareholders' meeting room and great hall which Philip Charles Hardwick had built for the London & North Western at Euston at a cost of around £150,000

(1846–9), arousing a storm of criticism from shareholders.[4]

Indeed, the 1840s had been a difficult decade altogether: the 'Hungry Forties'. Harvests had been poor for several years in a row causing widespread distress, and the catastrophe of the Irish potato famine of 1845–6 had seemed to mock Britain's much-vaunted political culture and scientific culture with their inability to solve the most basic of human problems, subsistence itself. The railway mania and the subsequent bust had been played out against a background of great political turbulence in Britain and Europe, culminating in the Chartist movement and the European revolutions of 1848. Regimes had fallen all over the continent, and were only precariously restored. Yet by 1850, with hindsight, a corner had been turned. Britain's economy experienced a major upswing: for the first time in history the four key indices of population, prices, GNP and wages, were all rising together. The plans for a great international exhibition of design and manufacture, planned by the Royal Society and announced at New Year 1850, were a harbinger of more optimistic times. The Great Exhibition and the Crystal Palace were important for many reasons, but among them, they were to be the key to the design of the new Paddington Station.

Over the second half of 1850 the GWR's board changed their mind; the formal decision to go ahead and build a new station at Paddington was taken on 21 December, by which time the plans were advancing rapidly. In their half-yearly report of February 1851 the directors said: 'It has been maturely considered whether any temporary additions to the present buildings at Paddington could have been usefully made, so as to postpone the construction of a permanent station, but every enquiry has proved that such expenditure would neither secure the main objects in view, nor be consistent with economy.'[5]

Figure 3.1
Bristol Temple Meads Station, from J C Bourne's History and Description of the Great Western Railway. *Note the carriage trucks in the foreground, the left-hand one with a carriage on it.*
[STEAM – Museum of the Great Western Railway, Swindon]

It was not just the Great Western who found that their first buildings soon became too cramped. The extreme case was Liverpool, where Crown Street Station had opened in 1830, to be replaced by Lime Street Station in 1836, which had to be rebuilt in 1846–51 and again in the 1870s.[6]

Most stations built in the 1840s were two-sided; that is, they had platforms and buildings on either side of the track, to serve the traffic going in each direction. When it came to planning a terminus, as at Paddington, this meant dividing the station between the departure side and the arrival side. However, for a terminus a third kind of plan, the 'head plan', was possible, in which the station was approached end (or 'head') on through a single concourse, opening into both the departure and the arrival sides.

Brunel's first real terminus was at Bristol Temple Meads, where the departure and arrival sides were kept separate, and the five lines of track between them ran through to the engine shed, which occupied the head of the station, a very unusual arrangement (Fig 3.1).[7] In his initial 1836 sketches for Paddington he had considered a different kind of plan, with a central roadway separating the trackbeds for the departure side and the arrival side. Since then, two superb head-planned stations had been started in Paris, Léonce Reynaud's first Gare du Nord (1845–7), and François Duquesney's Gare de l'Est (1847–52).[8] In both cases, the tracks

were all united in a single wide bed, with a departure platform and an arrival platform on opposite sides, accessed commonly from a concourse across the head of the station. The Gare de l'Est, which was and is particularly admired, still stands.

In 1850, across London, Lewis Cubitt was designing King's Cross Station for the Great Northern Railway, an elegantly planned compromise between the two-sided plan and the head plan. Unlike the French examples, the departure and arrival sides had separate track-beds under twin train sheds, given heroic expression in the giant arches of the south façade.

At Paddington, Brunel was limited in what he could do by his site, which lay well below street level. Late in 1850 he began to draw layouts for the new station in his sketchbooks, preserved at Bristol University Library (Fig 3.2). An initial sketch simply hints at a wide separation between the departure side and arrival side.[9] Two rather more developed sketches show a remarkable plan, unlike any previous station in Britain.[10] This picked up the theme of Brunel's unexecuted 1836 design with its central roadway. As then, he planned a grand entrance from Conduit (now Praed) Street. A roadway 58ft (17.7m) wide would have led down a ramp to a central concourse, about 125ft (38m) wide and 400ft (122m) long.[11] On either side are the tracks for the departure and arrival sides. The

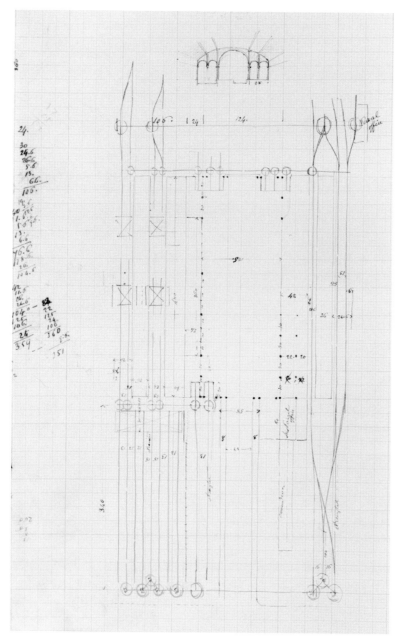

Figure 3.2
Brunel's sketch plan for a
new station at Paddington,
1850, with a thumbnail
design for the wide-span
roof he envisaged over the
central concourse.
[University of Bristol
Brunel Collection (Large
Sketchbook 3, 1). By
courtesy of the Brunel
Institute, a collaboration
of the SS Great Britain
Trust and the University of
Bristol]

spans on either side: it is not clear how these five roofs relate to the sketch plan. Brunel was experimenting with different designs for the openings in the end elevation: in one version it looks as though he is experimenting with horseshoe-shaped arches of an almost Moorish character.[12] The dimensions of this plan are not very clear, but the whole building seems to be 347ft (105.8m) wide.

Brunel clearly had an iron and glass roof in mind, but he had never actually built such a thing. His most ambitious roofs to date had been at Bristol Temple Meads (72ft (21.9m) wide) and Bath (50ft (15.2m) wide), both of timber and both of *c* 1840–1. That at Bath has disappeared without adequate record, but the original train shed at Bristol survives and has been studied in detail by John Binding.[13] There, Brunel started with a daringly experimental design: the roof would not be carried on conventional trusses at all, but on 22 pairs of crane-like cantilevered beams facing each other. Each crane balanced on a column as its fulcrum: the heavier outer section was tied back to the wall, while the long inner arm reached halfway across the shed: the point seems to have been that if the two cranes did not meet in the middle, they would not exert any outward thrust on the walls. Nothing like this had ever been built before and the problem seems to have been that a timber roof simply could not have the stiffness that Brunel's design required. It had to be modified during construction, with crownpieces and thick wrought-iron straps added to join each pair of cranes together, turning them into portal frames of a rather unconventional design. Brunel was still trying to stabilise the roof in 1849. So much ironwork was added that, as Mr Binding says, the roof could be described as a composite timber and iron structure, and despite these efforts the outward thrust generated by the roof has caused visible bowing in the columns and outer walls.[14]

The Bristol Temple Meads roof was a brave attempt to push engineering in timber rather further than it would really go. It had to all intents and purposes failed and required rescue, and by 1850 it must have looked like the deadest of dead ends compared to a number of wide-span roofs made of glass, iron and timber that had recently gone up. They represented a leap in building technology, which began with garden glasshouses and warehouses, and was further developed at a number of railway stations and naval dockyards.

more developed of these two plans shows the position of the stables and of a hotel.

In the 1836 scheme the central roadway seems to be left uncovered. However, the 1850 sketch plans appear to show the whole station roofed in. One of the sketches shows the column positions around the central concourse at 10ft (3m) intervals; another outline seems to represent the edge of the covered area, and some tiny vignettes show Brunel's first thoughts for a wide segmental-arched roof. Two sketches, close by in the same sketchbook, show an end elevation for the train shed. The central span has a broad segmental arch, with two smaller

In the 1830s a series of increasingly ambitious greenhouses were built at Chatsworth for the 6th Duke of Devonshire to designs by his head gardener Joseph Paxton. These culminated in the 'Great Stove' (1836–40), an iron-framed building 277ft (84.4m) long and 123ft (70m) wide, with a central nave 70ft (21.3m) wide spanned by laminated timber arches.[15] Around the same time a Dublin engineer and iron founder, Richard Turner, was also building superb glasshouses in Belfast (1839) and Dublin (1843). In 1844–8 Turner and the architect Decimus Burton collaborated on the Palm House at Kew Gardens, 362ft (110.3m) long, 100ft (30.5m) wide and 63ft (19.2m) high at the centre, at that date the finest iron and glass building in the world (Fig 3.3). Equally important in engineering terms, though less publicly known at the time, were a number of spectacular roofs over slipways in the naval dockyards at Portsmouth, Chatham and Pembroke. From around 1815 some very wide timber roofs had been built in the dockyards; in 1845–7 the Board of Admiralty had a series of magnificent iron roofs with spans of *c* 80ft (24m) built by two firms of contractors: George Baker & Son of Lambeth and Fox, Henderson & Co of Smethwick. The latter firm will reappear shortly as the builders of the new Paddington Station.[16]

In the mid-1840s railway planners saw the advantages of wide-span roofs that allowed them greater flexibility to rearrange the tracks and platforms beneath. The new building techniques, developed in glasshouses, began to be applied to railway stations. In 1845 the architect John Dobson designed the curved train shed at Newcastle upon Tyne with three parallel arched sheds each 60ft (18.3m) wide (Fig 3.4).[17] In the same year Eugène Flachat began the Entrepôt des Marais in Paris, with single-span arches 118ft (36m) wide. These were immediately eclipsed by the second Lime Street Station in Liverpool for the London & North Western Railway, begun in 1846 and with a single span roof 153ft (46.6m) wide and 374ft (114m) long. The Lime Street roof was made by Richard Turner, who was building the Kew Palm House at the same time: the responsibility for its design seems to have been shared by Turner and the London & North Western Railway's engineer Joseph Locke. It was carried on 'sickle' or 'crescent' trusses, in which arched ribs of wrought iron are tied together with wrought-iron tie-rods, braced apart by cast-iron struts, the whole forming a crescent-moon shape. This was pushing iron founding to new limits and the first experimental trusses failed when tested at Turner's foundry in Dublin. Lime Street Station was nevertheless completed in 1850 with the widest single-span roof in the world, wider than the Pantheon.[18]

Figure 3.3
Decimus Burton's Palm House at Kew Gardens of 1844–8.
[© Crown copyright.EH FF85/9]

Figure 3.4
Newcastle Central Station, by John Dobson and the engineering contractors Hawks Crawshay – an early photograph.
[Reproduced by permission of English Heritage BB81/3058]

In 1850 the new 'glasshouse' technology found even more spectacular expression in the Crystal Palace in Hyde Park, the most revolutionary building in the world. Brunel was closely involved in this as a member of the Building Committee for the Great Exhibition of 1851: the Crystal Palace is the key to understanding his rebuilding of Paddington (Fig 3.5).[19]

The idea for an 'Exhibition of the Manufactures of all Nations' came about in 1849 through the alliance of Henry Cole, Secretary to the Society of Arts, and the Prince Consort. At New Year 1850 a Royal Commission for the Exhibition was established under the prince's chairmanship. An Executive Committee was formed, driven by Cole and meeting every day. On 5 February 1850 the exhibition's Building Committee met for the first time. It numbered two aristocrats: the Duke of Buccleuch and the Earl of Ellesmere; three architects: Charles Barry, T L Donaldson and C R Cockerell; and three engineers: William Cubitt, Robert Stephenson and Brunel. They met about every four days for the next six months. It seems clear that the leading voices on the committee were Stephenson's and Brunel's.[20]

A competition was announced, and 230 entries were received in April 1850. They represented an extraordinary range of 'radiating' and 'parallel aisle' plans, built of iron and glass. Two were singled out for special praise, one by Richard Turner whose Palm House at Kew was then rising, and another by a French engineer, Hector Horeau. However, both were costed at around £300,000, as compared to the exhibition's available budget of £100,000. It was also a moot point how quickly they could be built: the Commission had already announced that the Great Exhibition would open on 1 May 1851. The Building Committee were alarmed. On 16 May 1850 they told the commissioners that they could not recommend any of the competition entries. They wrote a more detailed brief for the exhibition building, and set to work producing their own design. On 22 June this was published in *The Illustrated London News* (Fig 3.6). It was in effect a massive, brick-walled, triple-span railway shed; in the middle was a giant iron and glass dome 200ft (61m) in diameter, contributed by Brunel. There is something rather characteristic about the fact that Brunel, in his first attempt at designing an iron-framed roof, envisaged what would have been the widest single span in the world. The committee invited tenders from contractors, but the design caused a furore for its permanence and ugliness. It would have required 17 million bricks and there was much doubt as to how long it would take to build. By the first week of July the future of the Great Exhibition hung in the balance.[21]

However, Joseph Paxton, the Duke of Devonshire's brilliant head gardener had started to take an interest. Paxton was a self-taught genius, who had risen from a post at Kew Gardens to being the head gardener at Chatsworth, to designing and building glasshouses of increasing size and structural sophistication, and becoming the Duke of Devonshire's general adviser and right-hand man.[22] He was, among other things, the duke's representative on the board of the Midland Railway, and on Friday 7 June he was in London on railway business, to meet the Midland's chairman John Ellis MP at an acoustic test for the new Commons Chamber in the Palace of Westminster. The test was a disaster and Paxton expressed his concern of similarly fundamental mistakes being made with the exhibition building, telling Ellis that he 'had some thoughts of sending in a design that would solve the difficulties complained of'.

Ellis, knowing Paxton's talents, immediately took him to the Board of Trade to see Henry Cole, and be briefed on the exhibition's problems. He paced the site in Hyde Park, and then travelled to North Wales for the weekend, where he attended the floating and erection of the third of the immense 460ft (140.2m)-long tubes for Robert Stephenson's Britannia Bridge over the Menai Straits. Brunel was in London, working on his design for the exhibition building dome. On 12 June, the Commissioners for the exhibition invited tenders to build the Building Committee's design, due back in a month. Galvanised into action, Paxton and his staff spent nine days from 12 to 20 June making designs with help from the engineer W H Barlow. On Thursday 20 June he set out for London with his drawings, and happened to meet Robert Stephenson at Derby Station, returning from supervising work on Newcastle Central Station (which was to be the other great influence on Brunel's design for Paddington). They travelled to London together, Stephenson immediately recognising how superior Paxton's design was

to the Building Committee's own. On Saturday 22 June *The Illustrated London News* published the Building Committee's design, complete with Brunel's dome. There was an outcry in the press over its massive ugliness. Two days later Paxton met Charles Fox and Richard Chance of the iron founders Fox, Henderson & Company and the glassmakers Chance Brothers, both based at Smethwick in Staffordshire: this was first of several meetings to work out how his design could be manufactured and built in detail.[23]

Brunel was still standing by the official design, but he helped Paxton, advising him to change his design from a 20ft (6m)-module (ie, the standard span between the columns) to 24ft (7.3m), to comply better with the exhibition's brief. Paxton and Fox re-drew their designs. The elm trees in Hyde Park were another problem: they could not be felled, and Paxton was obliged to change his design again, introducing a great arched transept, in part for structural reasons, in part to accommodate the trees. As a gesture of friendship, Brunel turned up at Devonshire House (where Paxton was staying) early one morning with a note of the heights of the elm trees, and a note saying that he meant to win with their own plan, but 'I thought it right to give your beautiful design every possible advantage.'

On 10 July, the tenders were in for the official design. None of them came below the £100,000 limit: Fox, Henderson offered to build it for £141,000 for six months' use if they could retain the materials. Paxton's rival design, however, was by now published and known. Around 29 June Paxton had signed an agreement with Fox, Henderson & Co and Chance Brothers. Fox, Henderson would pay him £2,500 for his design, including the glass roofing system known as 'Paxton roofing' that he had developed the previous year for the Victoria Lily

Figure 3.6
An Illustrated London News *picture of the Building Committee's unsuccessful proposal for the exhibition building, 22 June 1850. [© Illustrated London News Ltd/Mary Evans]*

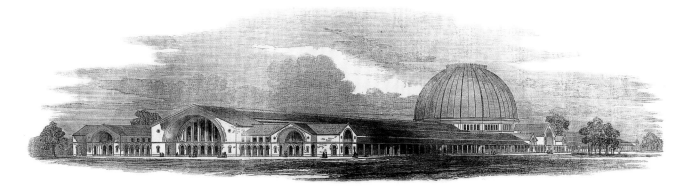

House at Chatsworth. Chance Brothers would supply the vast quantities of 16oz (453.6g) glass at 3d per foot and on 10 July Fox, Henderson & Company also offered to build the Paxton design for the remarkably low-seeming sum of £89,950, again retaining the materials afterwards. They guaranteed that the shell would be complete by 1 January 1851.[24]

There was a final crisis over the exhibition's lack of financial guarantees, solved by a generous intervention from the railway contractor Samuel Morton Peto, who provided a personal guarantee of £50,000. On 26 July the Commission finally decided, choosing Paxton's design and accepting Fox, Henderson's tender. The sequel was a miracle of organisation: as the historian John McKean has pointed out, what Paxton and Fox had designed in the Crystal Palace was as much a process or a system as a finished object. The vast building, 1,848ft (563.3m) long and 456ft (139m) wide, the largest enclosed space in Britain and probably the largest in the world, was built in less than five months. On 1 May 1851 the Great Exhibition opened as planned. It was the most remarkable and public demonstration of what British engineering was capable of, and one of the most extraordinary chapters in the history of construction.

The Building Committee's design had been well and truly defeated, but Brunel was too impressed by Paxton and Fox's work to feel resentful. He became an ardent supporter of Paxton's design, and as a member of the Building Committee was closely involved in its construction. Thus, while Brunel was planning the new Paddington Station in late 1850 and early 1851, he was also attending weekly meetings for the Great Exhibition and visiting the site in Hyde Park.

Brunel's first design for Paddington, as we have seen, was based on an entrance from Conduit Street leading down into a central concourse 450ft (137m) long and 125ft (38m) wide (see Fig 3.2). In December 1850, however, there was an abrupt change of design. This seems to have arisen from the board's decision to have a grand hotel at Paddington. The idea came from one of the directors, James St George Burke, whose letter urging the case for a hotel was read out at a board meeting on 21 September 1850.[25] The board were probably influenced by the fact that their new line from Oxford to Birmingham, then under construction, would put them in direct competition with

the London & North Western for the first time (the line opened in 1852). In 1839 the rival company had opened a pair of hotels at Euston, designed by Philip Charles Hardwick and framing his father Philip Hardwick's monumental Doric entrance.[26] Thus on 5 December the GWR board resolved to build a hotel too, observing that this amenity could 'decide passengers to adopt either one line or another', and setting a budget of £40,000 for it.[27] On 19 December 1850 Brunel attended another board meeting and 'submitted ground plans of an intended passenger station at Paddington such as he considers to be necessary, in order to determine the most eligible position for building the proposed Hotel'.[28]

The evidence of Brunel's sketchbooks suggests that he took a version of his 'central concourse' design to this meeting. One of the sketch plans, discussed earlier, has a rectangular block marked 'hotel' occupying part of the street frontage towards Conduit Street, much like his original scheme of 1836.[29] It would seem that the board rejected the central concourse scheme in order to give the whole of the Conduit Street frontage to the hotel, for it was only after this date that Brunel made the first sketch plans of the station as built.

Around this time Brunel chose T A Bertram, one of his assistants who had been the resident engineer for the eastern half of the main line, to help him with the designs for Paddington. On 20 and 21 December Brunel had meetings with Bertram 'concerning enlargement of Paddington Station'. For the next three months the office diaries record Brunel and Bertram having frequent meetings on the subject.[30]

On 21 December 1850 Brunel attended another board meeting with plans showing the 'future appropriations' of land that were needed for the goods yard and engine house. He was asked for estimates of the cost of the passenger station. The board then took the formal decision to go ahead. They did not want to make another temporary arrangement and felt it essential to decide on 'a permanent system of building', an interesting choice of words.[31] There was a problem, however, in that the existing goods station occupied part of the intended site. For this, as well as for financial reasons, the board ordered that work should start on the departure side (the south or Eastbourne Terrace side) and on the new goods facilities, so that there should be the minimum disruption of goods traffic.[32]

Three plans on consecutive pages of one of Brunel's sketchbooks represent the origin of the design for Paddington as built.[33] On page 11 there is a very approximate plan dated 26 December 1850, with a few key dimensions written in, all of them relating to the width of the building (Fig 3.7). The plan reverses the previous central concourse design in having the departure side and arrival side separated by a broad central trackbed. The cross-hatched rectangle at one end must represent the site of the hotel, greatly enlarged from the previous scheme.

On page 12 there is a much-developed but undated version of this. Many more dimensions are marked in. The train shed is marked as 750ft (229m) long (it is actually 700ft (213m)). The site for the hotel is shown in outline, and marked as 70ft (21m) deep. The area directly behind it is given over to the track-heads and turntables. The Eastbourne Terrace buildings are roughly marked in. Brunel has also marked out the curvature of the tracks as they bend southwards to go under the Bishop's Road Bridge.

On page 13 is a variation of page 12, also undated. Brunel has written several dimensions in but a lot have been crossed out and changed, and there are several columns of calculations: he was evidently thinking straight onto the paper (Fig 3.8). For example, he was trying to decide whether to make the Eastbourne Terrace buildings 30ft (9.1m) deep (page 12) or 22ft (6.7m) (page 13); this related to the decision whether the adjacent cab road should be 56, 58 or 60ft (17.1, 17.7 or 18.3m) wide. On page 13, he defines the site of the hotel as 290ft (88m) long by 70ft (21m) deep. The sketches are in pen but, wonderfully, on page 13 Brunel added the outline of three arched-span roofs in pencil as an afterthought. These swiftly scribbled lines are the origin of Paddington's famous roof. On New Year's Eve Brunel and Bertram spent five hours looking at Waterloo and Euston Stations.[34] On 2 January 1851 Brunel presented his revised plan to the board, which seems to date the three sketch plans to the last week of 1850.[35]

Brunel's initial scheme for a grand central entrance from Conduit Street, with departure and arrival platforms on either side of a great concourse, thus seems to have been scotched by the board's wish to provide a larger site for the hotel and to make it an architectural statement in its own right, rather than a subsidiary element of the station façade. Thus the station came to be concealed behind the Great Western Royal

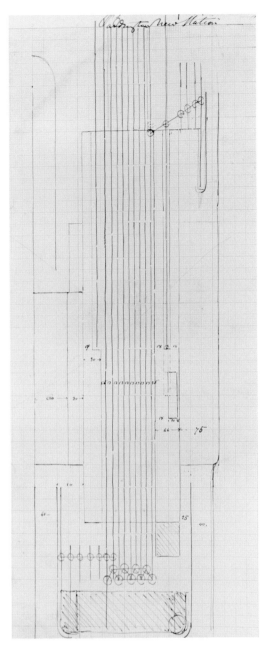

Figure 3.7
Page 11 of Brunel's 'Large Sketchbook 3', with the first of three successive sketches in which he developed the plan of Paddington Station. The rectangle at the bottom of the sketch represents the site of the hotel. [University of Bristol Brunel Collection (Large Sketchbook 3, 11). By courtesy of the Brunel Institute, a collaboration of the SS Great Britain Trust and the University of Bristol]

Hotel. Having started designing a sophisticated variation of the head plan with a central concourse, Brunel was forced back to a two-sided plan, with a broad central trackbed (Fig 3.9). It was, in fact, much like a larger version of his terminus at Bristol Temple Meads of 1839–41, with the hotel in the place of the Bristol office building, and the track-heads in the place of the Bristol engine shed.[36]

Over Christmas 1851 the board had engaged Philip Charles Hardwick as architect for the hotel (Fig 3.10). His father Philip Hardwick had designed Euston Station, originally intended as the GWR and the London & Birmingham

Figure 3.8
In the last of the three
sketches, the plan of the
station is largely worked out
as built. As an
afterthought, Brunel has
roughly drawn in the profile
of the celebrated triple-span
roof, again essentially as
built.
[University of Bristol
Brunel Collection (Large
Sketchbook 3, 13). By
courtesy of the Brunel
Institute, a collaboration
of the SS Great Britain
Trust and the University of
Bristol]

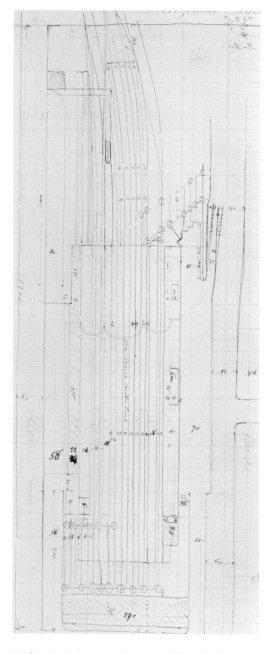

Figure 3.9 (opposite)
Paddington Station in 1854
(based on MacDermot
1964a, 172).
[© English Heritage]

for the departure platform and offices. The secretary, Charles Saunders, was ordered to raise with Brunel the desirability of 'having the general offices as concentrated as possible'.[39]

It seems curious that responsibility for the hotel should have been taken from Brunel and that he should apparently have acquiesced in this so easily. He was a very capable architect and had designed numerous buildings in a variety of styles for the Great Western Railway, including Bristol Temple Meads in a Tudor Gothic style, and Bath Station in Elizabethan style. Furthermore, he was generally given a very high degree of freedom and responsibility by the board. Had the hotel been left in his hands, one can readily envisage how he might have integrated it better with his station, perhaps retaining the essence of his central concourse scheme.

The answer may be that Brunel had to delegate responsibility for the hotel to Hardwick as he was simply too busy to design it himself. His office diaries for these years show a frantic schedule, with Brunel attending anything up to 10 different meetings a day, from seven in the morning until late at night, six days a week. In addition to designing Paddington, he was at this time working for the GWR's allied companies on the Oxford & Worcester Railway, the Birmingham & Oxford Junction Railway, the Vale of Neath Railway, the South Wales Railway, the West Cornwall Railway and the Chepstow bridge. He was also working for the Clifton Bridge Company and the Sunderland Docks Company. He had interests in the Galvanised Iron Company and the Vulcanized Rubber Company, and regularly attended meetings for them. He was attending meetings of the newly formed Eastern Steam Navigation Company (which was to build the *Great Eastern*). He was also regularly attending meetings of the Great Exhibition's Building Committee.[40]

It has been noted that Brunel was not in general a good delegator and that he had a marked tendency to try and control matters of detail, occasionally to the detriment of what he was doing.[41] That he was content to leave the hotel to Hardwick may reflect how hard pressed he was. This would seem borne out by his decision to engage some architectural assistance for the station itself. On 13 January 1851 Brunel wrote to the architect Matthew Digby Wyatt, who was secretary of the Executive Committee of the Great Exhibition (Fig 3.11). The letter is so important and so revealing as to bear being quoted in full:

Railway's joint terminus, and Hardwick Junior added the 'Adelaide' and 'Victoria' hotels (1839), and the great hall and shareholders' meeting room (1846–9).[37] He too attended the GWR's board meeting of 2 January 1851. It was held at Paddington, so they went to inspect the site of the hotel. Hardwick was commissioned to prepare plans and estimates: the hotel was to include rooms for third-class passengers and to house a subscription club.[38] He must have worked as fast as Brunel, for on 9 January he presented the board with a ground plan and elevation, which they approved on the same day as they accepted Brunel's general designs

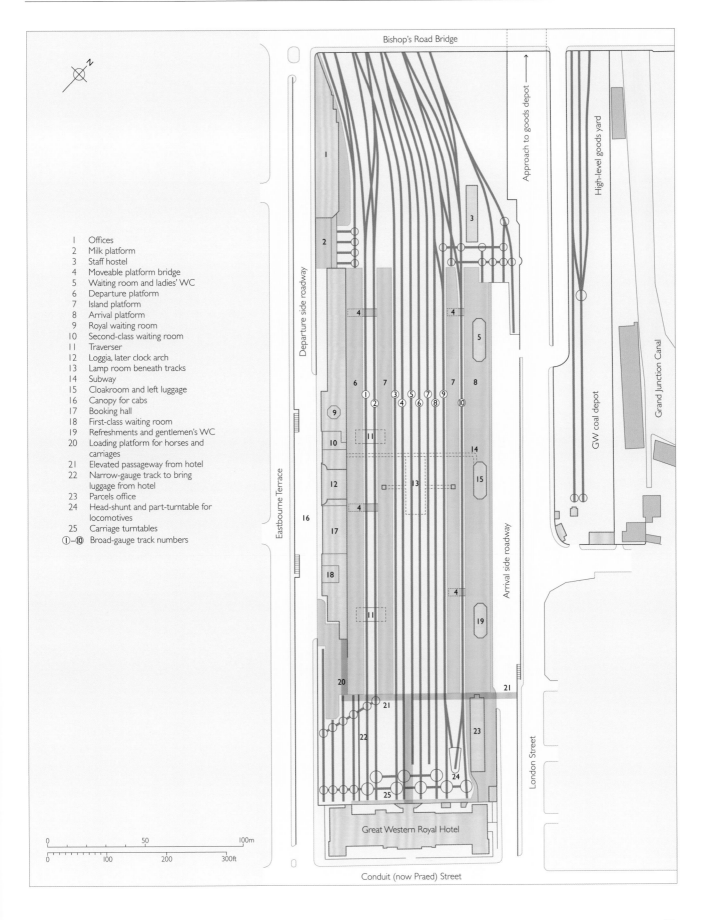

1   Offices
2   Milk platform
3   Staff hostel
4   Moveable platform bridge
5   Waiting room and ladies' WC
6   Departure platform
7   Island platform
8   Arrival platform
9   Royal waiting room
10 Second-class waiting room
11 Traverser
12 Loggia, later clock arch
13 Lamp room beneath tracks
14 Subway
15 Cloakroom and left luggage
16 Canopy for cabs
17 Booking hall
18 First-class waiting room
19 Refreshments and gentlemen's WC
20 Loading platform for horses and
     carriages
21 Elevated passageway from hotel
22 Narrow-gauge track to bring
     luggage from hotel
23 Parcels office
24 Head-shunt and part-turntable for
     locomotives
25 Carriage turntables
①–⑩ Broad-gauge track numbers

Bishop's Road Bridge

Approach to goods depot

High-level goods yard

Departure side roadway

Eastbourne Terrace

Arrival side roadway

GW coal depot

Grand Junction Canal

London Street

Great Western Royal Hotel

Conduit (now Praed) Street

Figure 3.10
Philip Charles Hardwick, a
watercolour perspective of
the design for the principal
façade of the Great Western
Royal Hotel.
[RIBA Library Drawings &
Archives Collections]

I am going to design, in a great hurry, and I believe to build, a Station after my own fancy; that is, with engineering roofs, etc. It is at Paddington, in a cutting, and admitting of no exterior, all interior and all roofed in … Now, such a thing will be entirely <u>metal</u> as to all the general forms, arrangements and design; it almost of necessity becomes an Engineering Work, but, to be honest, even if it were not, it is a branch of architecture of which I am fond, and of course, believe myself fully competent for; but for <u>detail</u> of ornamentation I neither have time nor knowledge, and with all my confidence in my own ability I have never any objection to advice and assistance even in

the department which I keep to myself, namely the general design.

Now in this building which, <u>entre nous</u>, will be one of the largest of its class, I want to carry out, strictly and fully, all those correct notions of the use of metal which I believe you and I share (except that I should carry them still further than you) and I think it will be a nice opportunity.

Are you willing to enter upon the work <u>professionally</u> in the subordinate capacity (I put in the least attractive form at first) of my Assistant for the ornamental details? Having put the question in the least elegant form, I would add that I should wish it very much, that I trust your knowledge of me would lead you to expect anything but a disagreeable mode of consulting you, and of using and acknowledging your assistance; and I would remind you that it may prove as good an opportunity as you are likely to have (unless it leads to others, and I hope better) of applying that principle you have lately advocated.

If you are disposed to accept my offer, can you be with me this evening at 9 ½ p.m.? It is the only time this week I can appoint, and the matter presses very much, the building must be half finished by the summer … Do not let your work for the Exhibition prevent you. You are an industrious man, and night work will suit me best.

I want to show the public also that <u>colours</u> ought to be used. I shall expect you at 9 ½ this evening.[42]

Brunel's invitation to M D Wyatt was not an isolated decision. He had already decided that he wanted the new station to be built by the great engineering contractors Fox, Henderson & Co, builders of the Crystal Palace. Later on Brunel followed up the hint in his letter to Wyatt about the use of colour by engaging the designer Owen Jones, who designed the colour scheme for the Crystal Palace, to do the same for Paddington. In effect, Brunel engaged the team who were building the Crystal Palace to build Paddington, with the obvious exception of Joseph Paxton. He may have consulted Paxton over the design, though this is not proven: Paxton was a director of the rival Midland Railway and thus could not have worked formally for the GWR.

As it happened, Matthew Digby Wyatt was not available to come and visit Brunel at half past nine on 13 January. Instead, Brunel saw Charles Fox of Fox, Henderson that evening (Fig 3.12).[43] This is the first meeting with Fox specifically recorded in the office diaries, though Brunel must have met him many times at Great Exhibition committee meetings. On 14 January Brunel had three meetings concerning Paddington. At 10 am he saw Bertram. At 5.40 pm, M D Wyatt came round in response to the letter quoted. At 8 pm, Brunel saw Charles Fox again. Over the next few weeks he had frequent meetings with Bertram and Wyatt concerning the station. On 22 January, a particularly hectic day, Brunel met Fox, Henderson again, and Bertram (twice).[44] On 23 January he presented the board with plans for the buildings facing Eastbourne Terrace and the southern of the three roof spans, as well as a schedule of prices from Fox, Henderson.[45] There is no evidence that any other contractors were invited to tender.

FRADELLE & MARSHALL. PHOTO.   230 & 246, REGENT S?

It was most unusual for the Great Western Railway, which was under great financial pressure, to accept a single tender from one contractor for such a large piece of work. Brunel must have explained this to the board as an exceptional measure, justified by Fox, Henderson & Company's outstanding abilities in iron construction, and in particular in view of their astonishing performance as builders of the Crystal Palace. His preference appears the more striking given that Sir Charles Fox was a well-known opponent of the broad gauge.

They were indeed a most remarkable company. Charles Fox, trained as a surgeon, had turned by the age of 20 to engineering, worked on the 'Rainhill Trials' of locomotives for the Liverpool & Manchester Railway and became a protégé of Robert Stephenson.[46] He worked on the London & Birmingham Railway, designing their extension into Euston and the train shed roofs there, which had the first all wrought-iron trusses made up of angles and tees (wrought-iron bars which had been run through a profiled rolling-mill to give them an L- or T-shaped section) (1836–9).[47]

Around 1839 he left the London & Birmingham Railway to form a partnership with the contractor and founder Francis Bramah, son of the famous iron founder and locksmith Joseph Bramah. They established a foundry at Smethwick near Birmingham, trading as Bramah, Fox & Co. In 1841 Bramah died, and Fox was joined by John Henderson, a Scots engineer. Fox, Henderson & Co were the first firm to produce the whole range of railway plant and equipment and also developed a capacity for major iron construction projects.[48] From 1843 the company won a series of contracts from the Royal Navy for iron structures in the royal dockyards, most notably five wide-span iron-framed roofs covering slipways, two at Pembroke (1844–5) and three at Woolwich (1846–7). It has been established that Fox, Henderson themselves designed the remarkable Pembroke slipway roofs, which had clear spans of over 80ft (24.4m), though other works were undertaken in partnership with the Royal Engineers.[49] Their business developed fast, and they added the GWR to their list of clients, supplying wrought-iron roofs 45ft (13.7m) and 55ft (16.8m) wide for the smiths' forges and the paint shop at Swindon in 1846–7.[50] By 1850 Fox, Henderson & Co employed between 800 and 1,200 men, their smiths' shop with its 70 forges was claimed to be the largest in the world, and they turned out about 300 tons (305 tonnes) of castings a week.[51]

Figure 3.13
A c 1914 view of Rewley Road Station, Oxford, built by Fox, Henderson & Co in 1851 for the Buckingham Railway.
[Reproduced by permission of English Heritage CC54/00324]

Through their work on Pembroke Dock and the Crystal Palace, Fox, Henderson developed great expertise in designing and building structures themselves, and for organising the logistics of large prefabricated buildings. They took on Brunel's contract for Paddington in January 1851 while still working on the Crystal Palace. At the same time they were building a smaller station for the Buckingham Railway at Oxford (Fig 3.13), of particular interest because its iron columns and beams were very similar to those of the Crystal Palace and because it survives to this day, albeit moved and reconstructed at the Buckinghamshire Railway Centre at Quainton.[52]

In the same year Fox, Henderson also contracted to build a new smithery in the Royal Dockyard at Portsmouth (October)[53] and the immense Birmingham New Street Station for the London & North Western. The last-named was comparable in size to Paddington, with the widest single-span roof anywhere in the world carried on huge crescent trusses 212ft (64.6m) wide (see Fig 3.23).[54] It is little wonder they became over-stretched. This seems to explain why their work on Paddington became so badly delayed, damaging their relations with the GWR as a result.

Throughout January, February and March Brunel was meeting Bertram two or three times a week to plan Paddington. Halfway through February another assistant, Charles Gainsford, was called in to help with the work. Brunel was certainly directing the effort but day to day the work was handled by his assistants. It is interesting to find that in 1889, from a GWR staff point of view, both the passenger and goods stations were remembered as having been 'built under the direction of Mr Gainsford, assisted by Mr W Jacomb, afterwards chief engineer of the GWR', without any reference to Brunel, Fox or Wyatt.[55]

Brunel met Charles Fox and M D Wyatt more irregularly. The office diaries, Brunel's sketch-books, and the drawings in Network Rail's archives between them give a fairly clear idea of how the responsibility for the work was divided. The whole of the planning was Brunel's work, as was the outline design of all the buildings except the hotel. The sketchbooks contain well-developed plans for the Eastbourne Terrace range of buildings (see Fig 7.18).[56] Brunel was responsible for the outline design of the train sheds, but his sketchbooks do not contain any detailed drawings for them. Elsewhere, the sketchbooks contain numerous details for roof structures, castings, girders and so on, but not for Paddington. The surviving detailed drawings for the roof structure were all produced by Fox, Henderson. That, after all, was the whole point of engaging them. A copy of the specification for their second contract, which was closely modelled on the first contract, has survived and provides detailed information about the materials used.[57]

Fox, Henderson were probably responsible for a crucial change to the shape of the arches. Brunel's sketches all show the train sheds spanned with segmental arches (see Fig 3.2): these would probably have needed tie-rods in order to be stable, like those at Newcastle upon Tyne, which would have interrupted the long vista.[58] Instead, the roof as built has beautiful five-centred arches, which give the design more visual 'lift' (Fig 3.14).[59] There may be a parallel here with Bristol Temple Meads: when it became clear that Brunel's original cantilevered timber roof would not work he, or one of his team, produced a design for strengthening it with tie-rods, but this was not carried out: the roof was strengthened in other ways, leaving its beautiful profile intact.[60]

Matthew Digby Wyatt's role, and the degree to which he shares Brunel's responsibility for the design, has attracted a good deal of scholarly attention.[61] Brunel may have delegated 'detail of ornamentation' in his letter to Wyatt, but this did not mean either that he disliked

Figure 3.14
Fox, Henderson & Co, a cross section of the station to accompany their first contract, 1851. Note the slender construction of the 'departure side' building on the left, and the timber platforms.
[Network Rail/EH AA031066]

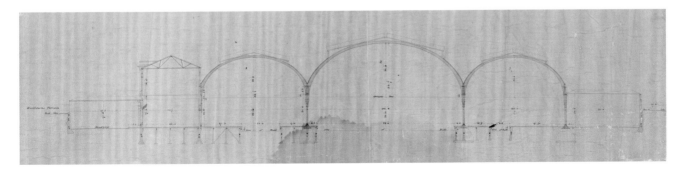

architectural ornament, or that he was incapable of designing it himself. Brunel's previous designs for the GWR demonstrate his clear understanding and judicious use of ornament: the handsome façade to the Box Tunnel may stand as an example (*see* Fig 1.8). Furthermore, the sketchbooks provide abundant evidence that Brunel had a half share in designing Paddington's remarkable decoration.

Pages 5 and 10 of 'Large Sketchbook 3', have sketches for the end façade of the first central concourse scheme. Brunel was experimenting with the segmental arch shapes for the shed roofs, framing semicircular or horseshoe arches carried on slender columns (*see* Fig 3.2).[62] On 7 January, by which time he had moved on to the revised plan, Brunel was again sketching architectural treatments for the end wall, generally similar in character. The same page, 22, has sketches for the long elevation of platform 1, one with round-headed windows, one with thin columns and flat lintels. Another tiny sketch shows columns and a girder, with a decorative pendant (Fig 3.15). On page 24 Brunel develops these ideas for the platform 1 façade, and for the column and lintel. This is the origin of the curious column capitals and of the decorative pendants hanging from the longitudinal girders (Fig 3.16). Page 28 has a swiftly dashed-off sketch for the platform 1 façade overlooking the two transepts (*see* Fig 7.9). On pages 33 and 34 are his first thoughts for the roof and railings of the departure side approach road, off Eastbourne Terrace. On page 37 is a cross section of the main train shed. Despite its fairly late position in the book, this still shows segmental-arch trusses, by this time with straight lines indicating the 'Paxton glazing' as used at the Crystal Palace. Brunel's sketches would have been developed into working drawings by Wyatt, who was assisted by Charles Fowler Junior.[63] Thirty-nine full-size detail drawings for ornamental cast iron for the roof are listed in the GWR's original catalogue of plans, but unfortunately they seem to have been lost.[64]

Matthew Digby Wyatt had actually designed very little before 1851, and at that point was perhaps more of an architectural writer and theorist than a practising architect. He had worked as junior partner to his elder brother Thomas Henry Wyatt *c* 1836–46, before embarking on a long tour of the continent in 1846–8, where he made great numbers of studies of historic buildings and stylistic details which were to provide the material for many books and lectures. After his return to England, he was commissioned by the Society of Arts to write an investigative report of an industrial exhibition in Paris in 1849: it was the quality of this report which led to his appointment, in 1850, as secretary to the committee of the Society which organised the Great Exhibition. In 1849–52 he was also the editor of the short-lived but influential *Journal of Design* (1849–52). Here he had made several contributions to the debate, then very current, about the use of new materials and technologies, and the role of ornament in architecture.[65] In 1849 Wyatt had reviewed John Ruskin's *Seven Lamps of Architecture*, the most influential work of artistic criticism in a generation. Ruskin's passionate hatred of industrial culture drove him into some contradictory attitudes. In 'The Lamp of Truth' he could write 'the time is probably near when a new system of architectural laws will be developed, entirely adapted to metallic construction', while adding, over the page 'the iron roofs and pillars of our railway stations are not architecture at all'.[66]

Wyatt admired Ruskin, and his review praised 'this thoughtful and eloquent book' for its 'eloquent denunciation of shams'. He criticised Ruskin, however, for his emotional resistance to industrial culture:

Instead of boldly recognising the tendencies of the age, which are inevitable, instead of considering the means of improving these tendencies, he either puts up his back against their future development, or would attempt to bring back the world of art to (what it was) four centuries ago … .[67]

In 1850 Wyatt had written another article for the *Journal of Art*, on 'iron work and the principles of its treatment'. In this, he wrote of Brunel:

His independence of meretricious and adventitious ornament is as great and as above prejudice as his engineering works are daring in conception and masterly in execution. From such beginnings, what glories may be in reserve, when England has systematised a scale of form and proportion, a vocabulary of its own, in which to speak to the world the language of its power … .[68]

So, Brunel and Wyatt both assumed that a building of Paddington's status required ornament: the point was to ensure that it was seemly and appropriate to its purpose. The 'correct notions of the use of metal' which Brunel referred to in his letter of 13 January to Wyatt can be deduced from the station itself. In the end elevations Wyatt substituted the beautiful tracery of vaguely Arabic shapes for Brunel's

Figure 3.15
Brunel's sketch designs for the new Paddington station – 7 January 1851. [University of Bristol Brunel Collection (Large Sketchbook 3, 22). By courtesy of the Brunel Institute, a collaboration of the SS Great Britain Trust and the University of Bristol]

Figure 3.16
An extract from Brunel's sketch designs for the new Paddington station – 9 January 1851. [University of Bristol Brunel Collection (Large Sketchbook 3, 24). By courtesy of the Brunel Institute, a collaboration of the SS Great Britain Trust and the University of Bristol]

horseshoe arches. For the column capitals he developed Brunel's sketch into geometrical figures, with no historic antecedents (Fig 3.17). For the arched trusses he devised the applied tracery, reminiscent of leaves, or of Gothic window tracery: it is possible that it plays a role in stiffening the arches. For the long platform 1 façade, he turned Brunel's sketch into a full design. The fat roll mouldings, rather appropriately, are made of cast-iron sections and on some of them the scratched foundry marks are still visible: they could literally be cast-iron gutter sections, pressed into ornamental use and embedded in a general layer of cement render. For the façades overlooking the two transepts, Wyatt worked Brunel's little sketch up into the design for the delicate oriel

windows, with their fantastic filigree and 'cut-out' ornament that we see today (*see* Figs 7.9 and 7.10).

Doubtless Wyatt had various historical sources for his contributions to Paddington, but the context and his handling of it effectively transforms it into something new, into what Henry-Russell Hitchcock called the 'High Victorian' style. There seems little point in pinning any historical label on Paddington other than 'Victorian', and that, presumably, was exactly what Brunel and Wyatt were aiming for.

On 17 April 1851 tenders for the hotel were opened: all were substantially higher than the board's estimate of £40,000. That of Messrs Holland, for £43,750, was accepted.[69] Fox, Henderson & Co, perhaps because they had taken on too many commitments and were over-extended, were slow with the first deliveries of ironwork. As early as 15 May 1851 came the first of a long series of complaints by the board about their tardiness and a request for Brunel to investigate.[70] On 28 May 1851 he was instructed to threaten Fox, Henderson with the withdrawal of their contract.[71]

The GWR needed to demolish the old goods depot, as it occupied part of the site for the new the passenger station. The new goods depot would go just to the north-west. Brunel started to think about it in January, with a cross-sectional sketch showing his idea for a 'transferring bridge from platform to warehouse'.[72] On 12 June 1851 he wrote to the board to say that the plans and specifications were ready (Fig 3.18).[73] Tenders for this were opened on 26 June 1851. They covered a disconcertingly wide range of sums, from £17,000 to £27,000, where Brunel's estimate had been £24,124.[74] Messrs Sherwood's tender, for £24,432, was accepted on 10 July.[75]

Work must have been proceeding on the departure side, but in October the board again complained of Fox, Henderson's 'insufficiency of means'.[76] On 20 November 1851 Brunel brought ground plans of goods buildings and warehouses.[77] On 27 November Fox, Henderson wrote to the board with a statement of progress, giving assurances that the existing contract would be completed by the end of March, as long as they were provided with the necessary drawings. Brunel stated that the outstanding drawings were for the gable end of the shed, the bow window (doubtless the windows overlooking the transepts of the train shed) and the entrance from the hotel.[78]

Figure 3.17
Paddington – a view along the arched trusses showing Matthew Digby Wyatt's decorative cast-iron tracery.
[© English Heritage K010980]

Right through the first half of 1851 Brunel's office diaries show that he attended several meetings a week, sometimes two or three in a day, concerning Paddington. Most of these were with his subordinates Gainsford and Bertram, but numerous meetings with Wyatt are recorded, and a few with Sir Charles Fox. On 3 February he had the first of several meetings with Sir William Armstrong, later Lord Armstrong of Cragside, the celebrated inventor, engineer, and manufacturer. Armstrong had pioneered the development of hydraulic power and its application to cranes and other pieces of heavy lifting machinery. The evidence for what happened at Paddington is patchy, but it suggests that Brunel had a vision of Paddington as a fully automated station, unlike any that had previously been built, with systems of traversers, dropping platforms, platform bridges and turntables, all manufactured by Armstrong's company at Elswick on the Tyne, and integrated into the design of his station to move goods, carriages, locomotives and wagons around in the most efficient way possible. This is discussed in more detail in chapter 5.[79]

On 27 November 1851 the board discussed the arrival side for the first time, and asked Brunel for a report and drawings. They instructed him to devise penalties to protect the company against further delays.[80] On 4 March 1852 the unfortunate Mr Wilder of Fox, Henderson appeared before the board. He promised that the approach road to the Eastbourne Terrace buildings would be completed in 10 days and that 400ft (122m) of slating (out of a total of 600ft) would be done in a fortnight. Work had begun on stuccoing the exterior.[81]

Despite the board's strictures about Fox, Henderson's slowness, they instructed Brunel in March 1852 to obtain a tender from the firm for making and building the rest of the main roof, to be completed within a given period for a fixed sum, subject to penalty clauses. Brunel presented summary estimates of the total cost of the work then in progress (the departure side, the goods station, the engine house, coal sidings and roadways, but not the hotel), of £221,350.[82]

At this point the makeshift organisation of work and the rather haphazard chain of command was graphically revealed. On 17 March 1852 Brunel wrote to Charles Saunders stating that he had been obliged to spend large sums on draughtsmen's work and on employing Matthew Digby Wyatt – £650 for drawings of the station in the previous quarter, and £624 to Wyatt for his fee and for his draughtsmen. Brunel had apparently neglected to inform the directors that he was employing Wyatt at all: he now attempted to argue that by making

Figure 3.18
One of Brunel's designs for the goods station showing details of the granary roof, c 1851.
[Network Rail/EH AA031051]

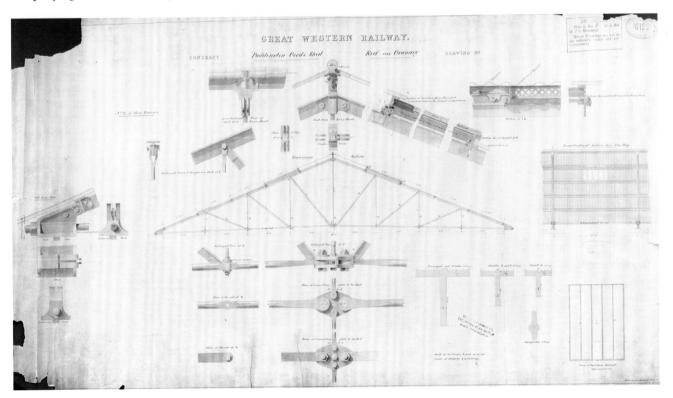

this informal arrangement he had saved the company money:

... if I had requested the Directors to call in an architect, as all other Companies have done for their stations, their charges would have been doubled and tripled, and I may safely refer to the Great Northern and the Euston Square and Camden Town stations as proofs that the building of railway stations by architects instead of engineers involves an excess of [illegible] in works which in our case would have not have been less than £20- or £30,000. The employment of an architectural assistant, entirely under my own control, I took upon myself ... .[83]

The board gave him £1,000 on 23 March but denied any responsibility for Wyatt's employment, or for any further remuneration:

The Board will appoint an architect, if absolutely necessary, to superintend buildings for stations etc., but they cannot agree to payment for the services of any architect employed by him to assist or advise him in designing portions of that work such as Mr Brunel admits to be comprised in the Agreement entered into for the three years, 1850–52.[84]

The clear implication was that the board did *not* accept that an architect was 'absolutely necessary', expecting Brunel to take care of this kind of thing himself. The directors had been kept in the dark about rather a lot. Apparently

Figure 3.19
Owen Jones, a portrait by H W Phillips. Jones, the leading theoretician of colour and ornament of the age, is shown against part of his 'Moorish Court' at the rebuilt Crystal Palace in Sydenham: Matthew Digby Wyatt drew on this kind of design from his friend Jones' Grammar of Ornament, *for the decoration of Paddington.*
[RIBA Library Drawings & Archives Collections]

they were also unaware that Matthew Digby Wyatt had consulted the designer and writer Owen Jones over colour schemes for the train shed, following up the hint in Brunel's original letter to him of 13 January 1851 (Fig 3.19).[85] Jones designed a rather elaborate scheme, apparently including decorative tiles as well as paint, and a trial area was carried out in April 1852, when the board came to inspect. They profoundly disapproved, and directed Charles Saunders to write to Brunel that 'The Directors ... beg you will not introduce any ornamental or coloured tiles, but will have the ordinary stucco ... under the roofs of the sheds. There is very decided objection on their part to any sort of decorative ornament to the passenger platforms or offices, which they wish to have as plain and inexpensive looking as possible.'[86]

No reference to Jones' work has been found in the GWR's records (presumably because he was paid by Brunel and Wyatt, not by the company). His involvement is confirmed by a magazine article about William Powell Frith's famous painting *The Railway Station*, in which reference was made to the directors having 'removed Mr Owen Jones' decorative colour from the roof and columns of the station'.[87]

Wyatt had also consulted the sculptor Alfred Stevens over the decoration of the octagonal royal waiting room, planned as part of the departure side buildings. Stevens produced a number of sketches for a magnificent decorative scheme in an Italian *cinquecento* style, which would have been immensely expensive (Fig 3.20). Again, there can have been no prospect of the board sanctioning this kind of thing.[88]

Messrs Holland had made swift progress on the great hotel, which was nearing completion in April 1852: part of it was apparently being used to house some of the passenger station's operations. A restaurateur, Monsieur Dethier, was negotiating to take a lease of it and the board considered whether to advance him a loan of £10,000 for the purchase of furniture; his surety was the Corsican hotelier Vantini who ran the Euston Station hotel and also the 'North Euston' at Fleetwood. On 27 May 1852 Hardwick submitted an inventory and estimates of furnishings needed: Dowbiggin, Holland & Co offered to furnish the hotel for £11,543. The board demanded a lower estimate, and the same firm obliged with a reduced version at £7,049 excluding carpets, curtains and bedding (3 June 1852). By October, though, Dethier had

Figure 3.20
Alfred Stevens, design for the
royal waiting room, 1853.
[RIBA Library Drawings &
Archives Collections]

withdrawn from negotiations and the board were once more looking for a tenant operator. At the end of the year they gave up trying to find a tenant and resolved to form a subsidiary company of their own to run the hotel, endowing it with capital of £15,000.[89] In February 1853 the hotel was said to be 'in a very forward state, and can be finished for occupation as soon as the Passenger business shall be transferred into adjoining buildings'.[90]

The total cost of building and furnishing the hotel was a comparatively modest £49,134.[91] Brunel himself chaired the hotel company from 1855 until his death. According to his son 'he found attendance at the meetings of the Directors and the supervision of the management of the hotel a very agreeable relaxation from the more important duties which took him to Paddington'.[92]

The first stage of the goods depot was nearing completion in June 1852. Mr Sherwood, the contractor, was summoned before the board on 25 June to explain delays, and vowed that it would be finished in two months. The same meeting resolved that all the conventional construction work to complete the passenger station should be let as a single contract, which was to be advertised for competitive tender. A second contract would be let for the rest of the train shed. A third contract was to be let for the engine house and other works for the locomotive department. Brunel submitted drawings for all of these works, which were approved. He was asked to set about getting 'the whole Passenger Station under construction' immediately.[93]

The decision to separate the earth moving and masonry construction from the iron construction was presumably influenced by the board's dissatisfaction with Fox, Henderson's performance to date. Nevertheless their services were retained, and on 1 July 1852 they offered to manufacture the ironwork for the rest of the station roof, including the 'Lawn' area behind the hotel, keeping their unit costs the same despite a rise in iron prices. The offer was accepted, and on 22 July the contract price was settled at £35,199, to be completed within six months. Oddly enough, their specification for the work is dated 8 November, over four months into the contract period, which may be evidence of Fox, Henderson's slowness.[94] At the same time the contract for the rest of the passenger station works (earth moving, foundations and masonry construction) was let to

Messrs Locke & Nesham for £10,998, and the contract for the engine house was let to the same firm for £14,130.[95]

Work presumably began on this next phase of construction in the late summer of 1852 but continued not to run smoothly; on 25 November 1852 the board were again discussing delays to the work. They found that Brunel had not been receiving regular reports from the clerk of works, Charles Gainsford. Henceforth he was to send regular reports to Brunel, and to themselves.[96]

Fox, Henderson's slowness remained a key theme in the directors' discussions. By the end of 1852 there had still been no deliveries of ironwork from Smethwick. Gainsford was sent to investigate, and on 17 February 1853 he reported that progress there was unsatisfactory.[97] The following week Sir Charles Fox came in person to explain the delays.[98] On 10 March 1853 Brunel attended a board meeting, bringing plans for the extension of the goods station; a second goods shed, similar to the first, coal drops, and facilities for stone traffic were to be added, at a further cost of over £50,000. At the same meeting Brunel submitted plans for the first-, second- and third-class booking offices, with tenders for furnishing them from Messrs Holland. The departure side was still not finished, however – a year after Fox, Henderson's promised completion date.[99]

Relations with Fox, Henderson reached crisis point in April. On 3 April there had still been no ironwork delivered, and work appeared to have stopped dead on the departure side as well. Brunel was said to be away pursuing 'other avocations' and the board were almost frantic with exasperation.[100] They instructed Saunders to write to Brunel on 4 April 1853, expressing their disappointment that they had not heard from him for some weeks, and demanding that he send an inspector to reside near the Smethwick works and make daily reports on Fox, Henderson's progress. Saunders concluded the letter with the fear that 'we shall again lose this summer the occupation of the new building'.[101]

The board's criticism of Brunel does not seem fair. They were surely aware that at this time he was supervising the construction of the GWR's allied lines, the South Wales Railway and the Cornwall Railway. In January 1853 he had awarded the contract for the Royal Albert Bridge at Saltash (Fig 3.21), also for the Cornwall Railway, and he was within a month of

completing the remarkable Wye Bridge at Chepstow for the GWR (opened fully on 18 April 1853). He was also designing the *Great Eastern*, the huge ship whose construction would dominate and dog the last years of his life.[102]

The directors had stronger words for their contractors. On the same day, 4 April, Saunders wrote to Fox, Henderson stating that the board 'felt compelled to … employ proper persons to enter upon and finish the work which has so long remained untouched and unfinished, within the new Buildings erected at Paddington for the General Offices and the Traffic Department at the intended Passenger Station'.[103] He threatened that the GWR would take work on the new roofs into its own hands as well if the contractors did not respond immediately.

This ultimatum seems to have been effective, for the work speeded up greatly. In August Brunel could report that:

… the works at Paddington Station for the passenger trains are still much in arrear, but are advancing towards completion. The alteration of the main lines … has been effected, and the engine house is in course of construction upon the site of the original line … As soon as it is completed, the old engine house may be removed and the extension of the goods department, so much needed, may be commenced.[104]

In August the size of the platforms was still being discussed. In October the decoration of the royal waiting room was ordered from Messrs Holland.[105] On 8 October 1853, *The Illustrated London News* could report that:

The vast works in progress at … Paddington are proceeding with very great rapidity … The new shed for the outgoing trains, being entirely roofed with glass, presents a light and handsome appearance, with separate entrance and reception rooms for the Queen and the Royal Family, and all the other requisite accommodation for the general public, which are of the most appropriate and complete character, are finished, and the foundation of the arrival sheds, with extensive offices, is now proceeding with great rapidity.

Figure 3.21
Brunel's Royal Albert Bridge at Saltash, under construction July 1858. The first span is in place and the second span has been floated out and has just begun to be raised.
[© Science Museum/Science and Society Picture Library]

On 27 October 1853 the board directed that traffic should be run out of the departure side from 27 December. Severe frost and snow prevented this and the first train was run out of the new station on 16 January 1854. Work on the main roof was nearing completion, but substantial elements of the arrival side remained to be done. In March 1854 Messrs Holland's offer to build the second-class waiting room on the arrival side for £987 was accepted.[106]

The board pressed for the arrival side to be opened on 18 May 1854. Brunel pressed for a delay of a week, as the roadway down to the platforms was still unpaved. In the event, the first train was run into the new arrival side on 29 May. The new Paddington Station was open and running, if not quite complete. The hotel opened on 9 June 1854, and was visited the following day by the Prince Consort and the Prince of Wales.

In January 1854, *The Builder* ran a short piece on the new station, remarking that 'Colour is sparingly introduced – red and grey. Each vault is covered at the sides, about half way up, with corrugated iron: the remainder is glazed.'[107] In June 1854 *The Builder* ran a longer article, with a plan and a full-page view of the train shed, saying:

The station may be called the joint design of Mr Brunel and Mr M D Wyatt, the former having arranged the general plan and all the engineering and business part ... the latter the architectural details in every department. The principle on which they set out was to avoid any recurrence to existing styles, and to try the experiment of designing everything in accordance with the structural purpose or nature of the materials employed – iron and cement.[108]

On 17 June *The Builder* carried more technical description, in particular of the train shed, with an elevation of part of the departure side façade overlooking one of the transepts.[109]

In July T A Bertram was appointed as the resident engineer.[110] The passenger station and the hotel may have been open, but to the west work was still in progress on many other departments. The northern part of the huge goods depot was finished, but the southern part and the high-level coal yard were still under construction, as were the staff hostel, the engine shed, the locomotive department offices and the locomotive superintendent's house (Fig 3.22). The wagon sheds, carriage

Figure 3.22
A house for the locomotive superintendent, Daniel Gooch, c 1853–5. The house stood near the engines at Westbourne Park. Demolished c 1906, its foundations were revealed during recent construction work there.
[Network Rail/EH AA031068]

sheds and stores were even further behind. All of these buildings have long since disappeared: a summary account of some of them is given in chapter 8.

It would be difficult to say at what point the rebuilding of Paddington was really finished – work continued on fitting out the station and its various dependencies right through 1857. By the end of that year the GWR had spend £668,790 on the new station, hotel, goods depot and associated works.[111]

A short postscript concerning the station's main authors seems in order. As work on Paddington neared completion Fox, Henderson & Co were at full stretch, building the 212ft (64.6m)-wide roof of Birmingham New Street Station for the London & North Western, the widest single-span roof in the world (Fig 3.23).[112] They were also rebuilding the Crystal Palace in a modified and permanent form, again designed with Joseph Paxton, in Sydenham. Paxton asked Brunel for advice about the 200ft (61m) water towers which were needed at either end, designed by the engineer Charles Hurd Wild, a request which led to the towers being demolished and rebuilt to Brunel's design in 1854–6.

Brunel recommended that Paxton employ Fox, Henderson to build the water towers as well as the Palace: 'With all their faults, and no man I believe has had much greater experience than I have of the faults of Messrs Fox Henderson as contractors, yet I am bound to state that if properly looked after they have the ability and the desire to execute excellent work.'[113]

As the delays in the construction of Paddington suggest, Fox, Henderson & Co were indeed becoming seriously overstretched. As railway building in Britain passed its peak, they like other firms were looking for work overseas and an ambitious contract to build the Zealand railway in Denmark proved their undoing. Fox, Henderson lost £70,000 on the job and became insolvent. In October 1856 the firm was declared bankrupt with liabilities of £360,000, causing 2,000 people to be thrown out of work.[114] Sir Charles Fox recovered from the crisis and went on to form a celebrated firm of consulting engineers in partnership with his sons. Sir Charles Fox & Sons later became part of Freeman Fox, now part of Hyder plc.[115]

Matthew Digby Wyatt revived his association with the Great Western Railway when they

Figure 3.23
Birmingham New Street Station, a print from The Illustrated London News *(3 June 1854).*
[Getty Images]

Figure 3.24
*Brunel at the launching of
the SS* Great Eastern, *Isle of
Dogs, 1857.*
*[© National Portrait
Gallery, London (detail)]*

conditions amongst the British wounded from the Crimean War. He responded even faster and more energetically than usual, producing designs and placing contracts for a 1,000-bed hospital in six days. A little over three weeks later, in mid-March, he had a prototype ward built at Paddington Station for the War Office to inspect. The resulting hospital opened at Renkioi on the Dardanelles in July 1855.[118]

More and more of his time was being taken up, however, by the *Great Eastern*, the largest ship that had ever been built, begun by John Scott-Russell and Company on the Isle of Dogs at the end of 1853. The building of the *Great Eastern* was immensely traumatic and difficult, and it wore Brunel out. Several attempts were made to launch it late in 1857 and 1858 (Fig 3.24). It was finally fitted out and ready to sail in August 1859. On 2 September Brunel collapsed of a stroke on board the ship. The *Great Eastern* sailed on its first voyage on 7 September. The same day there was an explosion on board. It did only minor damage to the ship, but the news of this probably hastened Brunel's death on 15 September at the age of 53. He was suffering from Bright's Disease or renal failure, for which there was then no treatment, which would have killed him in any event: we may speculate that he died prematurely, exhausted by his labours, but we cannot know this for certain.[119]

Daniel Gooch, his friend and colleague, wrote:

By his death, the greatest of England's engineers was lost, the man of the greatest originality of thought and power of execution, bold in his plans but right. The commercial world thought him extravagant, but although he was so, great things are not done by those who sit down and count the cost of every thought and act. He was a true and sincere friend, a man of the highest honour, and his loss was deplored by all who had the pleasure to know him.[120]

Brunel was succeeded as the GWR's chief engineer by T A Bertram, his deputy for the building of Paddington.

appointed him as architect for the reconstruction of Bristol Temple Meads station, *c* 1865–77, where he carried on the Tudor Gothic style established 30 years before by his friend Brunel. This time Wyatt was in partnership with Francis Fox, son of Sir Charles Fox.[116]

By the time the new station was complete Brunel was stepping back from some of his responsibilities as the GWR's engineer. He had already delegated responsibility for maintenance of the line to deputy engineers and in July 1854 he established a permanent engineering department at Paddington under Bertram to carry out this function. Brunel concentrated on new works, notably the great Royal Albert Bridge over the River Tamar at Saltash (1853–9).[117] Around 16 February 1855, Brunel was asked by the Government to help plan a prefabricated hospital building, to be sent to Turkey to alleviate the desperate

Maturity, 1855–1947

It is a tribute to Brunel's foresight and the quality of his design that it was 50 years before Paddington Station needed significant expansion, despite a tremendous growth in the GWR's network and passenger numbers. In the years after Paddington opened, the GWR experienced a generational change. Charles Russell, the chairman who had championed Brunel and the broad gauge, resigned in August 1855. He had been very highly regarded by the staff, and in 1852 1,700 of them had subscribed to the cost of a full-length portrait of him by Sir Francis Grant (Fig 4.1).[1] Always known simply as 'the Picture', it hung in the company's board-room at Paddington. For generations of GWR employees, being summoned before the directors was known as 'going to see the Picture'. Sadly, Russell's mental health was in decline and he died by his own hand in May 1856. In 1862 Charles Saunders, the company secretary who had done more than anyone apart from Brunel to create the GWR, resigned: he was 66 and suffering from heart trouble. By this time, only one of the directors had been on the board in the days of the gauge war in the 1840s. In 1864 Daniel Gooch, the locomotive superintendent since 1837, also resigned to take a part in the venture to lay a telegraph cable across the Atlantic. This seemed like a final break with the GWR's origins, but in 1865 Gooch returned as the company's chairman.

In 1860–2 Paddington itself became the subject of one of the most famous of all Victorian paintings. The artist William Powell Frith was looking for a suitable subject to follow his phenomenally successful crowd scenes, *At the Seaside* (or *Ramsgate Sands*) of 1854 and *Derby Day* of 1858. Realising that the great new railway stations represented meeting places where the whole of Victorian society could be seen, in 1860 he decided to paint a great panorama of Paddington, *The Railway Station* (Fig 4.2). Frith was already a celebrity, and there were

Figure 4.1
Sir Francis Grant's portrait of Charles Russell, chairman of the GWR (after it had been cut down in size). [STEAM – Museum of the Great Western Railway, Swindon]

regular reports on the progress of the work in the artistic press. In 1861 Frith pre-sold the painting with the reproduction rights for engravings to a major picture dealer, Louis Flatow, for the unprecedented sum of £4,500 (Flatow, a very skilled publicist, was later to claim that the sum was in fact 8,000 guineas). Flatow induced Frith to waive his right to display the picture at the Royal Academy for a further payment of £750, and it was instead the subject of a one-painting show at the Haymarket Gallery from April to September of 1862, visited by over 21,000 people in seven weeks. A short book was also provided, describing the action of the painting to the visitors. Flatow sold the painting with the reproduction rights to the publisher Henry Graves for the gigantic sum of £16,300, again to a flurry of press interest: Graves published a superb engraving of it by Francis Holl, which also proved very popular. In 1883 the original picture was bought by

Figure 4.2
The Railway Station *by*
William Powell Frith
(1819–1909), oil on canvas,
1862.
[© Royal Holloway and
Bedford New College,
Surrey, UK/The Bridgeman
Art Library RHC 2321]

the 'patent pill' magnate Thomas Holloway for just £2,000 and it remains in the gallery at the Royal Holloway College, Egham.[2]

The Railway Station, one of the great Victorian icons, is also a valuable record of Paddington's appearance when it was newly built. Frith had commissioned the photographer Samuel Fry to take a series of pictures of the station for him to work from, though the architectural background was actually painted by William Scott Morton. The painting looks diagonally across the train sheds from the departure platform, where a train is drawn up ready to leave behind one of Gooch's 'Iron Duke' class locomotives (labelled the *Great Britain*, though Frith's actual model was the *Sultan*). Over 60 figures crowd the canvas, including Frith, his wife and two of their children (as the family sending their sons off to boarding school, left of centre). Officers of the Metropolitan Police and the GWR's Railway Police are also represented. Frith accurately depicts the GWR porters' livery of dark green with scarlet stripes.

Frith commissioned a young artist, Marcus Stone, to paint two copies of the work, and one other copy seems to have been produced around this time. The GWR, naturally enough, were very interested in Frith's immortalisation of their station, though due to their financial plight they could not have contemplated buying it. They instead bought several copies of the engraving, and later on they were given one of Stone's copies: this was presented to George VI on the nationalisation of the company in 1947.[3]

The 1860s were very difficult years for the Great Western. Because they were no longer able to gain parliamentary permission for significant new stretches of broad-gauge track, they had to expand by buying hundreds of miles of standard-gauge lines in the West Midlands, which almost amounted to a separate railway system, based on Wolverhampton. To run efficiently they also had to extend their standard-gauge services southwards: hundreds of miles of track were thus converted to the mixed gauge on which both kinds of train could run, imposing a huge extra financial burden. The first standard-gauge train entered Paddington itself on 14 August 1861, an ominous moment for the company.[4]

Years of expansion and the need to build mixed-gauge lines had left the GWR with a huge floating debt of more than £13 million, the servicing of which was a great burden: of the major companies, probably only the London, Chatham & Dover was in a worse condition.[5] In 1858 profits dropped almost to nothing, and only some vigorous shareholder lobbying secured a dividend of 1.25 per cent, the lowest that the company ever paid. Worse was to come in May 1866, when the bank of Overend, Gurney & Company collapsed, causing a run on the banks and a general financial crisis. The GWR could no longer meet its debenture obligations. An attempt to raise £2.5 million in new shares was a fiasco and a request for a loan from the Bank of England was turned down. Sir Daniel Gooch, by now chairman of the GWR, approached the Government for support and was turned down there as well (Fig 4.3). In the first half of 1867 the company teetered on the verge of bankruptcy.[6] It recovered quickly, returning to profit in 1868, but the company remained short of money for several years to come. This, and the conservatism of Gooch's management, delayed full conversion to the standard-gauge for another generation.

Paddington has never been thought of particularly as a commuter station. For most of its history the GWR seems to have assigned a fairly low priority to suburban services, remaining true to Brunel's initial vision of the GWR as a long-distance network. In this, as in much else, they had much in common with their greatest rivals, the London & North Western Railway. Southern companies such as the London, Chatham & Dover, the London, Brighton & South Coast and the South Western Railway, knew that their long-distance business was never going to match the GWR's in scale, and probably for this reason they attached far more importance to developing suburban lines. This was a major factor in the development of south London, with its dense network of overground suburban lines fanning outwards from London Bridge, Cannon Street, Charing Cross and Waterloo, still the quintessential commuter stations. The Great Eastern Railway, hemmed in to its East Anglian territory, likewise spent much effort developing its network of suburban services: the historian Robert Thorne has written that 'when the Great Eastern built Liverpool Street it had its suburban services most in mind', and showed that these services and the Great Eastern's deliberate policy of serving working-class people were important formative influences on the development of east London.[7]

The GWR opened what might today be described as suburban lines to Windsor (1849), High Wycombe (1854), Uxbridge (1856), Henley (1857), Marlow (1873) and Staines (1885), but they hardly compare to the great networks in south London and it is worth considering why. The GWR certainly knew that there would be a demand for what we now call commuter services and by 1865 they had introduced season tickets between Paddington or Farringdon and stations as far as Windsor, Reading, Hungerford and Oxford: a season ticket from Paddington to Ealing cost £10 in first class, £7 10s in second.

During the railway mania of the 1840s and into the 1850s the GWR was at full stretch,

Figure 4.3
Sir Daniel Gooch.
[© National Railway Museum/Science and Society Picture Library]

running broad and mixed-gauge lines and staking out its territory as far and as fast as possible. By the time the new Paddington was opened in 1855 the company was paying the price for this hectic expansion with huge debts and a falling share price. Despite its difficulties, the GWR invested in the Metropolitan Railway when it was established in 1852. The Metropolitan was the first true underground railway in the world and in the 1860s it formed the basis of the GWR's suburban services.

There was a story, or myth, that Brunel had built Paddington in a cutting because he foresaw the building of underground railways.[8] This may seem far-fetched and is most unlikely to have influenced his choice of site in 1836. Paddington nevertheless turned out to be at just the right level to be linked to the underground Metropolitan Railway: on 2 November Brunel attended his first meeting on the subject, rather tellingly referred to in the office diary as 'City Extension to Paddington'.[9] This is interesting, as this was well before the Great Western Railway had become formally involved in the Metropolitan: perhaps Brunel was investigating on the GWR's behalf.

The Metropolitan did not obtain its Act of Parliament until 1853 and then it had trouble raising money. A sharp lawyer, Acton Smee Ayrton, joined the board: later, as First Commissioner of Works in Gladstone's administration of 1868–74, he was to be notorious for his insistence on economy and his bullying of architects. He persuaded the Great Western to invest £175,000 by proposing to link the new railway to the GWR lines at Paddington. Opposition in Parliament and financial difficulties delayed matters and work did not begin until February 1860.[10] The GWR's large shareholding meant that they could insist on the laying of mixed-gauge track and the linkage of their lines just north-west of the new Paddington train shed (Figs 4.4 and 4.5). Here they built the new Bishop's Road Station, which was the Metropolitan's original western terminus.[11]

When the new railway opened in January 1863 it was worked by the Great Western with broad-gauge rolling stock. This arrangement did not last long, but the new line enabled the GWR to run services through to Farringdon (1863) and later to Moorgate (1866).[12] The GWR's early relations with the Metropolitan were fairly stormy but nevertheless opened the way for them to develop what might be described as a small suburban network by proxy. The Great Western and the Metropolitan both bought shares in the Hammersmith & City

Railway (now the Hammersmith & City Line), which opened in June 1864, again worked by the GWR. Trains could now run from the Metropolitan Line through the new Bishop's Road Station at the north-west corner of Paddington, along the GWR's tracks to Westbourne Park, and then 2½ miles (4km) south to Hammersmith Broadway, thus winning traffic from the new suburbs of Notting Hill, Shepherd's Bush and Hammersmith.[13]

Another of the Hammersmith & City's attractions was that it could be linked to the West London Railway, till then one of the less successful legacies of the 1840s. The West London, which ran south to Kensington from the junction of the GWR and the London & North Western Railway at Willesden, was the only section of the ambitiously titled Birmingham, Bristol & Thames Junction Railway to have been built. It had opened in 1844, but its passenger services were so unsuccessful that they closed a few months later, and the GWR and the London & North Western Railway took a joint lease of the moribund line the following year. It was little use to them, carrying

Figure 4.4
Paddington Station and its environs from the first edition 25in. Ordnance Survey map.
[© The British Library Board (OS 25" (1869) XXXIII)]

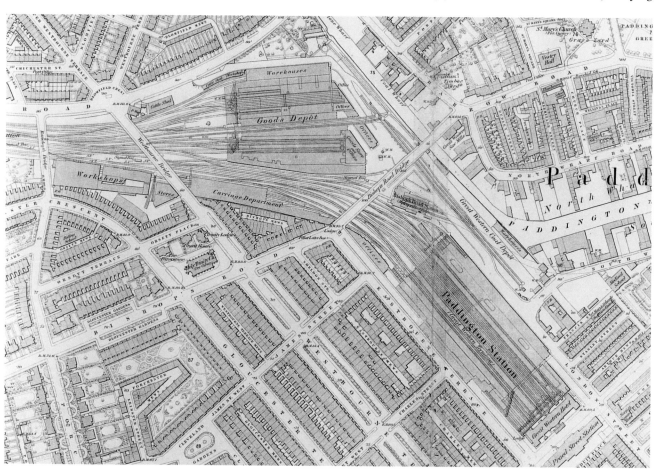

only the occasional goods train. In 1858 the bill for a railway to Victoria Station was passed and the GWR and London & North Western were spurred on to try and make better use of the West London. They each put in £100,000 and secured £50,000 from both the London, Brighton & South Coast and the South Western Railways and obtained an Act to extend the West London south through West Brompton and Chelsea, over the river to Clapham Junction. From being a cul-de-sac it was now a link between the lines of four great companies. The extension opened in March 1863, and in April the GWR began running broad-gauge services into Victoria Station, where a new wooden terminus had been rushed up.[14]

In 1864 the Metropolitan Railway was extended from Paddington to South Kensington. The new branch headed southwards just to the east of Conduit (now Praed) Street and involved the construction of a new station opposite the Great Western Royal Hotel, designed by Sir John Fowler, the Metropolitan Railway's engineer, and opened in 1867.[15] Like the Metropolitan's other stations it was built in a cutting lined with massive retaining walls of yellow stock brick and covered with an iron and glass shed roof. Praed Street Station (as it now is) was linked to Paddington by a subway in 1877. Today, it forms Paddington's Circle Line platforms (Fig 4.6; *see also* Fig 8.2).

Between 1852 and 1866 the company had been brilliantly inventive, making strategic alliances and investments and drawing in other companies' money, so they could run broad-gauge trains to the City of London, to Hammersmith, Clapham Junction and Victoria. The GWR only ran broad-gauge trains on the Metropolitan Railway until 1869.[16] In every other respect the policy nevertheless seemed to be successful and a report of the same year on suburban traffic towards Reading and High Wycombe showed that passenger numbers had risen from 513,018 in 1861 to 859,849 in 1867.[17]

Figure 4.5
'Junction of the Metropolitan and Great Western Railways at Paddington', an engraving from The Illustrated London News.
[City of Westminster Archives Centre]

Figure 4.6
The Metropolitan Railway Station on Praed Street (formerly Conduit Street), opposite the Great Western Royal Hotel.
[City of Westminster Archives Centre]

The GWR also ran a 'Middle Circle' service from Moorgate, via Earl's Court to Mansion House, around what we would now call the Circle Line, from 1872 until the Metropolitan Railway was electrified in 1906. In addition they rather briefly ran suburban services over lines belonging to the London & South Western Railway to Richmond in 1870, and again after 1894.[18]

One may wonder why the company did not do more to develop suburban services after their initial burst of activity in the 1860s. The answer may be to do with their financial difficulties: the GWR was already under severe financial strain, and the Overend & Gurney banking crash of 1866 brought the company to the brink of insolvency. The GWR recovered slowly: this and the heavy financial burdens of building the Severn Tunnel and converting the network to the standard gauge kept it short of cash for the next 20 years or so.

The GWR built two more lines in the London suburban area, though it is doubtful whether either of them can really be regarded as suburban lines in the true sense. In 1860 they opened an extension from their main line at Southall to the dock at Brentford on the Thames, but this seems to have been intended primarily for goods. Much later, the company built a new line from Ealing, via Greenford, to High Wycombe, in partnership with the Great Central Railway. This opened in 1904–5, and would probably be more accurately regarded as part of a new direct line to Birmingham than as a suburban line.[19]

By 1900 much of the traffic of the western suburbs had been picked up by the Metropolitan District Railway, which ran its lines from South Kensington to Hammersmith (1874), Richmond (1877) and Ealing Broadway (1879). Later, the Central Line ran out to Shepherd's Bush (1900) and Ealing Broadway (1920). By this time it was too late: Paddington would never be a great suburban terminus and the limited business it had was coped with by the relatively small Bishop's Road Station.

Brunel's great shed proved large enough to absorb half a century's growth in traffic. As railways became part of national life, however, so they changed society and generated new demands for travel. Railway excursions were born, and the GWR found that their business for August Bank Holiday week rose by 20 per cent in the five years between 1876 and 1881.[20] More platforms had therefore to be added, filling up

Brunel's broad central trackbed, in 1873, and in 1883, and another in 1889.[21] Between 1878 and 1882 a new 'milk platform' was added at the north-east corner of the train shed.[22] A couple of other alterations only require brief mention: a first-class waiting room added to the arrival side in 1875 has long gone.[23] The subway beneath Praed Street to the Metropolitan station, opened in 1877, remains as the access to the Circle Line platforms.[24]

The company's network and its bureaucracy were growing steadily and in the 1870s and 1880s the GWR had to expand their office accommodation. A proposal of 1872 to fill the gap between the Eastbourne Terrace offices and the hotel was not executed,[25] but in 1876 the gap between the offices and the isolated wing to the west was filled in, at a cost of around £28,369. This made space for the finance, engineers', estate and solicitors' departments.[26] In 1878–80, a further programme of rebuilding and enlargement was carried out. A second storey and a loft were added to the whole length of the range: the big skylight over Brunel's booking hall was sacrificed and built over, in the interests of making more room (Fig 4.7).[27] The Eastbourne Terrace range was extended eastwards over the carriage entrance to the departure platform. Beyond this, going towards the hotel, another nine-bay extension was built by Kirk & Randall of Woolwich for a comparatively lavish £37,827.[28] This paid for a richer architectural treatment, with a facing of Greenmoor stone, window aedicules and a bracketed cornice. Only half the building remains, the rest having been destroyed in the Second World War. A further scheme of 1881, to build more offices with a 60ft by 40ft (18m by 12m) shareholders' room and a 100ft (30m) tower was abandoned for financial reasons.[29] Meanwhile, inside the station, the track-heads and turntables in the area now called the Lawn were curtailed, providing a wide paved area for horse-traffic.[30] All of these works seem to have been designed in house by Edmund Olander, the GWR's engineer and Lancaster Owen, their architect.[31]

Still there was not enough space. The GWR took over several houses on the opposite side of Eastbourne Terrace as overflow office space: by the time of nationalisation in 1947 they had occupied most of the terrace.

In December 1880 electric light was installed at Paddington, one of the very first permanent installations anywhere in the world. In the years

Figure 4.7
An early 20th-century view
of the south flank of the
station towards Eastbourne
Terrace with the GWR's
offices.
[© National Railway
Museum/Science and
Society Picture Library]

around 1880, electric light had developed from an entertaining laboratory experiment to a practicable alternative to gaslight: the very first proper installation seems to have been a run of arc lights along the Embankment in 1878. Within the GWR the initiative came from the chairman, Sir Daniel Gooch. He asked his chief engineer, Dean, and telegraph superintendent, Spagnoletti, to report on the options, and they reported to the board on 10 November. The Brush system seemed to be the best: they tabled a plan for 34 arc lights for the passenger station and a quotation from the Anglo-American Electric Light Company for £1,953. Another £447 would be needed for a boiler and a shed to house the dynamo. With the brisk decisiveness of Victorian industry at its best, the directors agreed and the system was in by Christmas.[32]

These early arc lights were too bright for domestic use; they were also smoky and not entirely reliable. The inventor Joseph Swan had just invented the incandescent filament light bulb. This was first installed by Sir William Armstrong (later Lord Armstrong) at his house, Cragside, in Northumberland in December 1880, but it was some while before it was commercially available.[33] The GWR's arc-light system was dogged by difficulties, and in February 1884 the company signed a contract with the Telegraph Construction & Maintenance Company, who offered to re-equip and supply the whole of the station and offices for three

years for £4,200 per annum. A generator house was converted out of a carriage shed behind Gloucester Crescent. Two 90ft (27m) chimneys were built. Two dynamos, each weighing 4½ tons (4.6 tonnes), were installed, with a third in reserve: the magnet wheels were 9ft 8in. (2.9m) in diameter. Each dynamo was driven by a pair of tandem compound engines of 600 horsepower. Three smaller Crompton dynamos, powered by Willans engines, supplied current for the magnets of the main machines. The system supplied 4,115 of the new incandescent lamps in the offices and hotel and 98 arc lamps in the passenger and goods stations. It was in operation by February 1885, but immediately there were bitter complaints from the neighbours about the noise, vibration and smoke. Injunctions were served by the Bishop of London's estate, the dynamo had to be shut down pending experiments to dull the noise and the GWR and the Telegraph Construction & Maintenance Company got into contractual difficulties.[34] The new system may have caused problems, but this was probably due in part to its pioneering status. *The Electrician* described it as the 'first effort to be made in England to give a continuous supply of electric light on a very large scale' (Fig 4.8).[35] The system was replaced and a new generating station built out at Park Royal in 1907.[36]

In the 1870s the Great Western Railway recovered financially, thanks to Sir

Figure 4.8
The GWR's dynamo room in
1886 (from The Electrician,
28 May 1886).

Daniel Gooch's careful stewardship and a general upturn in the economy. It merged with its old partners the Bristol & Exeter Railway (1876), the South Devon Railway (1878), the West Cornwall Railway (1878) and the Cornwall Railway (1889).[37] It had the second largest network of any British railway (after the London & North Western Railway), and had returned to steady profitability, but the broad gauge, its most distinctive feature, now looked like an anomaly. The company had mixed-gauge lines across most of its network, and a majority of its services were now standard-gauge trains. Substantial sections of the network were converted in 1872 and 1874 and by 1878 only 7 of 48 daily departures were of broad-gauge trains.[38] Sir Daniel Gooch resisted conversion of Brunel's trunk route, and only after his death in 1889 did the company brace itself for this break with the past. The last broad-gauge through-train to Penzance, the Cornishman Express pulled by the locomotive *Great Western*, left Paddington at 10.15 am on Friday 20 May 1892 (Fig 4.9). Broad-gauge trains left for Plymouth at 1 pm and 3 pm, and at 5 pm the last broad-gauge departure left for Bristol pulled by the *Bulkeley*, which returned to Paddington with the 'up mail'. At 5.30 am on Saturday 21 May, the *Bulkeley* steamed back to Swindon with an empty train, the last broad-gauge journey of all, and over the following two days the remaining 171 miles (275km) of track were converted to the standard gauge.[39]

It was the end of a great engineering vision, but the broad gauge had been holding the GWR back, and the immense costs of gradual conversion had been diverting resources away from improvements to service. The company's reputation for high speed was a thing of the past. The slowness of many of its services in the 1870s and 1880s had saddled 'God's Wonderful Railway' with the derisive tag 'the Great Way Round'.[40] From 1892, the company entered a phase of vigorous innovation, which its historian E T MacDermot called 'the Great Awakening', and the period 1892–1914 is often referred to as the GWR's golden age.[41] In 1902 a new locomotive superintendent, George Jackson Churchward, was appointed. Over 20 years he was to develop engines of almost unrivalled quality, in the 'City', 'Saint' and 'Star' classes. Churchward's work improved the GWR's standard locomotives too; he replaced the old clerestory coaches with a more modern design, lit the coaches with electricity, and experimented with rail-motors, a petrol-electric railcar, and a steam-powered 'auto-train' which could be driven from either end.[42] Meanwhile a new traffic superintendent, N J Burlinson, was raising the speed of the GWR's services generally. The Welsh coal trade came into its own, with millions of tons pouring through the Severn Tunnel every year. The main line to Devon and Cornwall was thronged with new holiday services and an inspired slogan, 'the holiday

Figure 4.9
The end of the broad gauge:
the 10.15 am Cornishman
Express pulled by the
locomotive Great Western
leaves for Penzance, 20 May
1892.
[Great Western Trust
Collection]

line', sketched out a new image for the GWR that was later developed in hundreds of specially commissioned posters.[43]

By the 20th century, Brunel's great train shed was used to capacity. The south or departure side of the station, with the cab road and Eastbourne Terrace immediately beyond, was lined entirely with offices and could only have been enlarged with great difficulty and at immense cost. The GWR's traffic was growing fast. In 1856 Bradshaw's guide had listed 20 trains leaving Paddington each day; by 1907 this had risen to 300 every 24 hours, not including excursions, empty stock and 'specials'.

Around 1903 the GWR's Traffic Committee began to plan accordingly. The 'up relief' line was extended into the train shed as a stop-gap measure.[44] In June 1904 the general manager

placed a plan for major alterations to Paddington costed at £96,950 before the Traffic Committee, which was swiftly authorised by the board.[45] It was designed in-house, probably by William Armstrong, the GWR's new-works engineer. The plan provided three more tracks and platforms, each over 800ft (244m) long, two for passenger trains and one for milk traffic. A second line would be laid alongside the existing platform 9. All this would require radical reconstruction, including the digging out of London Street, the construction of a massive new substructure, and the addition of a new fourth span to the main train shed (Fig 4.10). New footbridges would link the enlarged station with the Bishop's Road Station (the present Metropolitan Line and Hammersmith & City Line platforms). On 8 August 1906

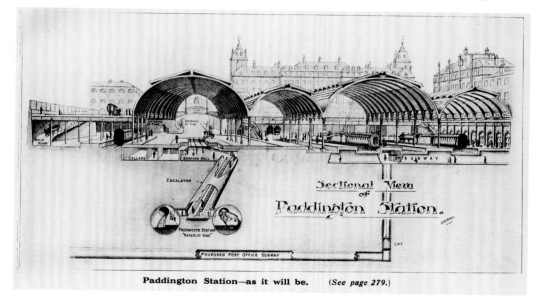

Paddington Station—as it will be. (See page 279.)

Figure 4.10
'Paddington Station as it
will be', a 1906 cutaway
view showing the planned
arrival side
improvements.
[Great Western Trust
Collection]

the general manager presented revised plans costed at £115,500.[46] These were approved by the board the following day, a testimony to the GWR's efficient management style.

Several phases of enabling works were required. The first major step was the building of a huge new engine shed at Old Oak Common, 3 miles (4.8km) out from the terminus: this opened in March 1906.[47] In 1906–7 the Bishop's Road Bridge, Westbourne Bridge, Ranelagh Bridge and Lord Hill's Bridge (all of 1835–7) were replaced with long-span steel girder structures to allow greater flexibility in planning the permanent ways beneath.[48] At about the same time new goods offices were built flanking the Bishop's Road Bridge to the south, a long four-storey block, 220ft (67m) long and 40ft (12m) deep, in a Victorian Classical manner which stylistically could have been built in the 1870s.[49] A new stationery department was built on Porchester Road at the south end of Lord Hill's Bridge with a reinforced concrete frame, faced in red brick with patent Victoria stone dressings. The building also housed a new electric sub-station for the Metropolitan Railway.[50]

In 1908 the main departure platform (now platform 1) was lengthened westwards to provide a new milk platform. The main thrust of the new work, though, was on the arrival side. In 1908 the old high-level coal depot between the station and Paddington Basin was closed, together with London Street, the oddly named thoroughfare running along the north side of the passenger station. Brunel's arrival side and the near end of the Mint stables were demolished, the whole area excavated down to platform level, and a massive new concrete retaining wall built towards the canal.[51] In April 1909 a contract for £22,500 was let to Messrs Jackaman of Slough for erecting the steel columns and superstructure to carry the re-routed London Street.[52]

A major element of the work was rebuilding the near (south-west) wing of the Mint stables as the new 'up' parcels depot. The site was dug down by another 7ft (2.1m) to platform level to allow a large basement to be built of reinforced concrete on the same Hennebique system as had been used on the Porchester Road building. François Hennebique had developed his ideas for using steel reinforcing bars to give tensile strength to concrete in France and Belgium in the 1890s.[53] By 1900 several patented systems were in use, but Hennebique's became the most successful, partly thanks to his marketing

techniques and flexible licensing arrangements. The first British application of it was at Messrs Weaver & Co's provender mill in Swansea in 1897.[54] Hennebique's principal agent in Britain was the French engineer Louis-Gustave Mouchel. L G Mouchel & Partners designed numerous buildings using the Hennebique system in Britain and also arranged for the system to be licensed to other contractors.[55] The GWR were one of the first licensees and their engineers' department was designing buildings using the system from c 1905–6.[56] Most early examples of the Hennebique system have been demolished and the basement of the Mint stables is a notable survival.[57] At platform level, a large circulation area to receive parcels and a sorting area were made; lifts linked these to a road depot at street level, where six loading docks allowed the parcels to be put straight on to the GWR's wagons.[58]

A contract for renewing part of the high-level coal viaduct was also let in August 1909.[59] There followed further contracts for excavating the site of the new goods and cab ramps (December 1910),[60] the diversion of London Street (February 1912),[61] construction of a retaining wall (April 1912),[62] the new cab ramp and cellarage (January 1913),[63] and widening the arrival side approach (April 1913).[64] In November 1913 the contract for the new steel roof over platforms 9, 10 and 11, the fourth span, was let to Messrs Holliday & Greenwood, for £28,969.[65] As a preparatory measure the northern row of Brunel's cast-iron columns was replaced in riveted steel.[66] The new roof was also designed by the GWR's engineering department. It was closely modelled on Brunel's three spans but wider than any of them at 109ft (33m). The steelwork was manufactured by the Horseley Bridge & Engineering Company (Fig 4.11; see also Fig 7.26).[67]

The new roof was nearing completion in June 1915. At this point the GWR board asked their engineer and new-works engineer to report on the train sheds generally. This was prompted by the collapse of Charing Cross Station's roof (designed by Sir John Hawkshaw and dating from 1863). The engineers did not think that the Brunel roof was in any danger but assumed that at some point it would be renewed. They noted, though, that the northern row of Brunel's columns was being pushed out of true because of the difference in width between his old north span and the new fourth span. They recommended adding tie-rods to the fourth span to remedy this.[68] They further recommended that the remaining original cast-iron columns be replaced in steel 'as a stage in the direction of the ultimate complete renewal of the roof'.[69] The existing octagonal columns of riveted steel were installed in 1922–3.[70] The new arrival side was brought into use gradually: the new platform 13 from November 1913, platforms 11 and 12 from December 1915, and platform 10 early in 1916.

The GWR was heavily engaged in the First World War, though not in so visible a way as the South Eastern & Chatham Railway, which took all the troop trains from Victoria to the Channel ports (Fig 4.12). The GWR hauled coal from Wales to feed the industrial war effort and 25,000 of the company's servants (about a third of the total) were called to the colours. At the same time, surprisingly, the company had to cope with booming passenger numbers. Britain was not mobilised for war in 1914–18 to anything like the same extent that it was in 1939–45 and it is strange to reflect that, in the terrible summer of 1916 while the battle of the Somme was raging, three Cornish Riviera trains carrying a total of 1,400 people were leaving Paddington every day.[71]

During the First World War, 2,524 GWR employees lost their lives. The company

Figure 4.11
The fourth span under construction, c 1915.
[Network Rail]

commissioned the very fine memorial on platform 1, designed by Thomas Tait, with the life-size bronze figure of an infantryman sculpted by Charles Sargeant Jagger. It was dedicated by Viscount Churchill, the then chairman, on Armistice Day 1922 (*see* Fig 7.11).[72]

In August 1914 the Government had taken overall control of Britain's railways, in order to manage war mobilisation. This control outlived

Figure 4.12
The Forces' canteen at Paddington, 1919. Note the abundance of flowers, the portrait of the king-emperor and the absence of alcoholic refreshment.
[STEAM – Museum of the Great Western Railway, Swindon]

the war and led to the Railways Act of 1921, which merged all the railway companies in Britain into four big groups with effect from 1 January 1923.[73] The Great Western Railway was the least affected of all the companies, being the only one to retain its original name and to maintain its identity largely undiluted.[74] It absorbed a whole family of railway lines and docks in South Wales, tied together by the GWR's main line, that added only 560 route miles (900km), but another 18,000 employees. This cemented the company's dominance of the Welsh coal trade, although that trade, like many of the Welsh coalfields themselves, was already in decline. The enlarged GWR, in addition to being the only major company to preserve its essential corporate identity undiluted, was also the only one of the 'Big Four' to have just one London terminus: the newly formed London Midland and Scottish, by contrast, had Marylebone, Euston and St Pancras. So Paddington retained its dominant position as the company's headquarters, as well as its pivotal place in the network.

The war caused great disruption to the railway industry and the decade 1910–20 turned out to be the last period when it could generally count on rising revenues. In the 1920s road traffic grew steadily, and the railway industry stopped growing. Nevertheless, for the time being the GWR seemed to recover strongly, thanks to Felix C Pole, their brilliant general manager from 1921–9. In 1923 the company returned an 8 per cent dividend, the highest in 75 years.[75] At Paddington, Brunel's goods depot was filled to capacity, with 900 road vehicles passing through it a day and 900,000 tons (914,44 2 tonnes) of general merchandise being handled each year. The goods depot was demolished in stages during 1925 and 1926. The 1854–8 buildings were demolished and replaced with a huge single shed measuring 625ft (191m) by 353ft (108m), with a seven-span steel-framed roof covering five double tracks (Figs 4.13 and 4.14; *see also* Figs 8.8 and 8.10). In the 1920s the company were still using horses for all their short-haul goods traffic in London, but a motor transport department was

Figure 4.13
Demolition of Brunel's goods depot, July 1925.
[© National Railway Museum/Science and Society Picture Library]

Figure 4.14
Goods are transferred from
the old goods depot to the
new, March 1926: note the
ubiquity and variety of hats.
[© National Railway
Museum/Science and
Society Picture Library]

being built up and in 1932 the company opened a depot for their 500 lorries at Westbourne Park.[76] The GWR were determined to compete with the road lobby on its own terms, but this was nevertheless an ominous sign for the future.

The General Strike of 1926 gave the industry a bad jolt (Paddington was kept running through the strike by 2,234 volunteers) (Figs 4.15 and 4.16).[77] Throughout the decade the GWR weathered the storm well. It had the benefit of an outstanding general manager in Sir Felix Pole, a latter-day Saunders, and in 1922 the great G J Churchward was succeeded as locomotive superintendent by C B Collett, who equalled Gooch and Churchward's achievements by producing the 'Castle' and 'King' classes, among the finest steam locomotives that have ever been built.[78] Thanks to them the Cheltenham Spa Express ran from Swindon to Paddington at 71.4 miles (114.3km) per hour in 1932, the fastest scheduled train service in the world.[79]

Figure 4.15
Some well-bred lady
volunteers, working in the
Mint stables during the
General Strike, 1926 (their
footwear rather gives them
away).
[© PRO Rail]

Although the GWR's business had peaked in the 1920s, Paddington was still under great pressure. In 1926 the general manager commissioned a report on its future, and T S Todd of the engineer's department produced a remarkable document that set out a master plan for doubling the size of the station. The Great Western Hotel would have gone, Eastbourne Terrace with all its houses would have been cleared, the site would have been dug out and another three roof-spans added to match the existing four. The architect Sir Henry Tanner produced some rather bland Classical designs in 1928, reminiscent of his work on Regent Street, for the immensely long façades which would be created on Praed Street and Westbourne Park Terrace (Fig 4.17). The new hotel, to the south, would have had a faintly American quality, like Tanner's Park Lane Hotel but on a bigger scale.[80]

It was a vision of Brunellian scale and thoroughness, but it came too late. The General Strike of 1926 had represented a bad knock for the company. The last good year was 1929, when the GWR returned a dividend of 7.5 per cent, the highest amongst the 'Big Four' railway companies. The Great Depression of 1929–31 hit all the companies hard and their revenues

fell alarmingly: for the railway industry as a whole, it would never be glad confident morning again.[81] The GWR was still capable of innovation, introducing colour light signalling on the main line from Paddington to Southall in 1932, and diesel railcars to British railways in 1933–4. It was also the first British railway company to introduce automatic train control, but in the mid-1930s it began to withdraw from the technological race: the kudos of having the fastest engines passed to Sir Nigel Gresley at the London & North Eastern Railway.

In 1930 the GWR embarked on another major round of improvements with the help of £1,000,000 in Government loans under the Development (Loans, Guarantees and Grants) Act of 1929, but it was a scaled-down project, in keeping with the straitened times.[82] More powerful locomotives meant that longer trains were now in use; a major element in the works was the lengthening of the main platforms, with lower shelter-roofs extending beyond the train sheds. The other main objectives were to modernise the Great Western Royal Hotel, to build new offices for the GWR, and to link the two sides of the station together more firmly. This last objective was to be achieved by converting the area at the back of the hotel, long

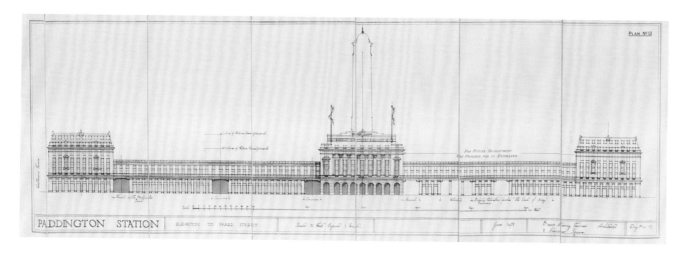

PADDINGTON STATION

known as the Lawn, then in use as a vehicle and loading area, into a pedestrian concourse.[83]

The works were designed by P A Culverhouse, the GWR's architect, and Raymond Carpmael, their chief engineer, and carried out between 1930 and 1935. Culverhouse and Carpmael were also responsible for a major remodelling of Bristol Temple Meads carried out at the same time.[84] As in 1906, there were extensive enabling works. To free the Lawn of vehicles, a new parcels depot was needed and a large new steel-framed building went up for the purpose on Bishop's Road with a separate access to platform 1. The hotel was given its own vehicle delivery entrance off the arrival ramp (at the north end of the hotel). To the south of the Lawn, there had remained a gap between the offices to Eastbourne Terrace and the south end of the hotel. Here, Culverhouse designed a new block of offices, steel-framed, eight bays wide and six storeys high, that filled about half the gap in a very restrained Classical style, again faced in patent Victoria stone. This was built c 1931–3 by Holliday & Greenwood.[85] The remaining gap between this building and the hotel was filled a few years later by a similar steel-framed building housing yet more offices, built by A D Dawney & Co.[86]

North of the Lawn, Culverhouse cleared the low 19th-century buildings which flanked the arrival side roadway.[87] Here he built another steel-framed office block in a distinctive Art Deco style, with shallow oriel windows incorporating ventilation vents. The name 'G.W.R. PADDINGTON' marched in bold letters across the top of the building and Culverhouse designed shell-shaped uplighters projecting from the façade, to illuminate it (Fig 4.18).[88] On the east side of the Lawn the hotel was extended, with a Post Office sorting depot on the ground

floor, offices for the publicity department on a mezzanine, a new drawing room for the hotel above that, and new bedrooms higher still.[89] At platform level, these new buildings had regular, simple Classical façades of patent Victoria stone facing onto the Lawn, which was paved as a proper concourse. Finally, Culverhouse and Carpmael covered the Lawn with a new steel-framed glass roof, with laylights around the perimeter, glazed in white and amber glass (Fig 4.19).[90] One result of these works, from the passengers' point of view, was to dispel the sense of Paddington having distinct departure and arrival sides, though the terminology survived in use among the station's staff into the 1980s.

Figure 4.17
The 1928 master plan for rebuilding Paddington, with immensely long Classical façades by Sir Henry Tanner and a giant new hotel facing Westbourne Terrace. This was overcome by the Depression and Culverhouse's more modest plans were carried out instead.
[Network Rail/EH AA031078]

Figure 4.18
The new arrival side offices by P A Culverhouse (from Anon 1933a, the GWR Magazine, March 1933).
[STEAM – Museum of the Great Western Railway, Swindon]

Figure 4.19
The Lawn roof newly rebuilt,
1933.
[Network Rail]

At the same time, Carpmael and Culverhouse gave the Great Western Royal Hotel a general overhaul, the first since its construction in the 1850s. Victorian architecture was at its critical nadir and Hardwick's grand interiors, including his magnificent two-storey coffee room, were destroyed. A series of low-key Art Deco interiors were created instead, with an American bar and a number of function rooms, lined with polished flush panelling throughout. On the station side, a new lounge was added above the new buildings and overlooking the Lawn. Measuring 87ft (27m) by 34ft (10.5m), it was decorated in orange and cream colours with bronze light fittings. Hardwick's hotel façade was subjected to a similar process of simplification. The cast-iron porch was removed and the two lower storeys refaced in patent Victoria stone in a much plainer style, with simple banding and plain window surrounds.[91]

The waiting rooms on platform 1 were redecorated, panelled in oak with ebony inlay. Platform 1 also received a 'Quick Lunch and Snack Bar', which sounds rather like a sign of the times, with its horseshoe-shaped bar surrounded by modish 'green-topped chromium-plated stools'.[92] At the north-west corner of the station, huge groundworks were carried out to widen the tracks, involving a partial rebuilding of the Bishop's Road Bridge.[93] The old Bishop's Road Station was demolished, the tracks widened, the name dropped, and the area redesignated as platforms 13 to 16 of Paddington Station proper.[94] A new booking office was built on the bridge-link to the main station.[95] The permanent way was remodelled for three-quarters of a mile (1.2km) out from the station.[96] Culverhouse and Carpmael also designed a large new building for the stationery department, with hostel accommodation for the hotel and refreshment department, on a quarter acre (0.1ha) site between Westbourne Terrace and the railway tracks.

Unknown to Culverhouse and Carpmael as they were modernising Paddington in 1935, the station acquired a place in architectural history in a quite different way. Two American architects and brothers-in-law, Nathaniel Alexander Owings and Louis Skidmore met in London and over refreshments at Paddington made a 'formal partnership' deal 'to offer a multidisciplinary service competent to design and build the multiplicity of shelters needed for man's habitat'. Thus Skidmore Owings Merrill, ultimately one of the largest and most influential architectural practices in the world, was born.[97]

By the time the modernisation of Paddington was complete, the company's business had recovered slightly from the depth of the Depression of 1929–31. The GWR had developed a fleet of lorries to supplement its horse-drawn van services; it had also set up bus services, though in 1928 it had sold a half-share in them: evidently it was responding vigorously to the challenges of the road lobby, and these would have gone some way to offset the fall in railway revenues, but in 1932 the dividend fell to 3 per cent, and the future was looking difficult.[98] Nevertheless, the company's centenary in 1935 was celebrated on a grand scale, with re-enactments of the first company meeting and the departure of the first train from Paddington, and with distinctive new coaches for the Cornish Riviera Express.

In 1938 the 'Big Four' launched a 'square deal' campaign to appeal for Government support against the rising road transport lobby.[99] If there was a bright spot anywhere in view, it was the continued prosperity of the company's holiday services, such as the Cornish Riviera Express and its special excursion trains, such as those for the Henley Regatta (Figs 4.20 and 4.21). Paddington was the pre-eminent 'holiday' station and the GWR had never been more popular, its image carefully polished by its skilful marketing department. The popular nostalgic image of the GWR, of gleaming green locomotives with burnished copper funnels pulling chocolate-and-cream carriages, of beautiful posters advertising West Country resorts, jigsaw puzzles of GWR engines, and immaculately uniformed inspectors, is largely a recollection of the company in the 1920s and 1930s, its silver age. This image was warmly evoked by the comic novelist P G Wodehouse in his novel *Uncle Fred in the Springtime*, published in 1939. Lord Ickenham, on his way to Blandings Castle (by the 2.45 express – Paddington to Market Blandings, first stop Oxford), reflects on the station's dignified character:

… there is something very soothing in the note of refined calm which Paddington strikes. At Waterloo, all is hustle and bustle, and the society tends to be mixed. Here a leisured peace prevails, and you get only the best people – cultured men accustomed to basset hounds, and women in tailored suits who look like horses. Note the chap next door. No doubt some son of the ruling classes, returning after a quiet jaunt in London to his huntin', shootin', and fishin'.[100]

One may imagine that the ghosts of the first GWR directors would have been suitably gratified.

Figure 4.20 (left) Holidaymakers at Paddington, 30 July 1945. [© Planet News/Science and Society Picture Library]

Figure 4.21 (below) Crowds travelling to the Henley Regatta, 2 July 1908. [© National Railway Museum/Science and Society Picture Library]

Figure 4.22 (right)
Women waving farewell
to soldiers departing from
Paddington, 1942.
[Getty Images]

Figure 4.23 (below)
Police and VAD (Voluntary
Aid Detachment) nurses
shepherd a crowd of young
evacuees, Paddington,
18 June 1940.
[Getty Images]

The GWR went to war together with the nation in 1939 (Figs 4.22 and 4.23). This time, total mobilisation meant that its resources were all at the state's disposal. Of the 15,000 of the company's servants who served in the forces, over 600 were killed or reported missing in action. Another 68 were killed by enemy action while in the company's service at home.[101] The head office left for Beenham Grange in Berkshire, which was probably just as well for Paddington was badly scarred by bombing during the war, the passenger station being hit seven times and the goods station four times. Two bombs caused especially serious damage: on 17 April 1941 a parachute mine hit the East-bourne Terrace buildings, demolishing a large part of the range just to the right of the platform 1 entrance together with the canopy over the cab road (Fig 4.24). On 22 March 1944 a pair of 500lb (227kg) bombs came through the train shed roof, blowing a great crater in the middle of the tracks (Fig 4.25).

Bombing and the demands of war placed great strains on the station and its staff. The writer Sir Osbert Sitwell left an evocation of this, having been summoned by Queen Mary to keep her company at Badminton House, Gloucestershire, for Christmas 1944:

The station-master's real office had been bombed, so the staff were using temporarily the royal wait-ing-room, a pretty octagonal room, crowned with a shallow dome, and having on one wall a rather pretty plaster medallion of Queen Victoria as a young woman. There were three men and two girls working, and the sound of feet outside was accompanied within by the tapping and click of typewriters, and the con-tinual ringing of a telephone bell. The station-master himself often rang up a mysterious higher authority, active elsewhere. This power he approached rather as a mortal might attempt to speak to those who dwell on Olympus. 'Is Mr Williams there ? … I want to speak to Mr Williams Himself. It's the Station-Master. Yes, I said HIMSELF … Is that you Mr Williams? The lawns are dangerously crowded, sir. I thought you might care to inspect them, and pass your opinion. You may think the gates should be shut.' (It was the first time I had heard platforms referred to as *lawns*, and this interested me. It must be, I think, a relic of days when platforms were open strips of grass).

At this moment, attention was diverted, for from his olympian retreat the great Mr Williams gave the order to shut the gates of the station, and this was done, so that now two crowds roared and raved, one inside, the other outside and besieging the yellow bulk of Paddington … From the office, the noise they made sounded like a rough sea … .[102]

Figure 4.24
Paddington at war: a
parachute mine hit the
Eastbourne Terrace
buildings on 17 April 1941.
[© PRO Rail]

The Great Western Railway coped well throughout the war, despite the demands of total mobilisation, bomb damage and the shortage of investment. After victory in 1945, though, the economic and political climate looked bleak. Even before the war the 'Big Four' had been suffering from the sharp rise in motor transport and their revenues were falling.[103] State control and the demands of wartime service had sapped their independence and profitability. The deciding factor, though, was the Labour Party's landslide victory in the 1945 election. The Attlee Government came to power on a manifesto of sweeping nationalisation that included the railway network. The GWR would have faced a difficult future whatever happened, but it was still a well-managed and profitable business, returning a final dividend on its ordinary shares in its last year of 7.23 per cent. The company's achievements had been tremendous, it was universally admired and respected, and it may well have faced its problems with all the vigour it had displayed in previous crises, but it was not allowed to: its fate was determined by politicians, not by economics. The British Railways Act received royal assent on 6 August 1947 and the vesting date was set as 1 January 1948.[104] On New Year's Eve the company handed over 100,000 employees,

3,856 steam locomotives, and over 3,800 route miles (6,100km) of track, to the new British Railways Board. The last official act, the final shareholders' meeting, took place on 5 March 1948 in the hotel at Paddington, not because it had business to transact, but from a widespread feeling that the meeting should go ahead anyway, and with that the Great Western Railway Company ceased to exist.[105]

Figure 4.25
A direct hit. On 22 March
1944, two 500lb (227kg)
bombs fell through Brunel's
roof and exploded on the
tracks.
[© PRO Rail]

The Station at Work, 1855–1947

In the Victorian age, the great railway companies were almost uniquely complex organisations. The only other bodies that could remotely be compared to them in their combination of large-scale human, financial and technological organisation and wide geographical reach were the Army, the East India Company and the Royal Navy with its dockyards, communications and victualling branches. For the whole of its history until nationalisation in 1947 Paddington was the hub of an industrial empire and was in itself a community the size of a small town. The artist William Powell Frith intuitively grasped that Paddington, like the Great Western Railway, could be seen as a microcosm of Victorian society (*see* Fig 4.2). Few historians have tried to set the railways in their social and economic context and no one has yet written a social history of the GWR.[1] A work of this kind,

taking in the company's staff, its services, and their impact on the communities they served, could be a lifetime's study: it is not possible in a book of this scope to accomplish very much in this direction. The aim is merely to set out some of the main themes and give a general idea of how the station operated.

Brunel designed the whole Great Western Railway as an operating system, and in the same kind of way he designed Paddington Station around the movement of trains and passengers. The three pages of his sketchbook, on which the plan of the station took shape in the last week of 1850, are essentially track diagrams, with the outlines of buildings sketched in around them (*see* Figs 3.7 and 3.8).

It is usual in planning a station to provide platforms on either side of the tracks, one platform for each direction. In planning a terminus, this translates into departure and arrival platforms, or in railwaymen's parlance 'down' and 'up'. In the 19th century it was unusual for a train to come in to a terminus, drop off passengers, pick up others, and set out again (what are termed 'in and out' services): trains did not run frequently enough for this to be manageable. Instead, all trains arriving at a terminus would be disassembled, cleaned and serviced, and new trains put together to be shunted into the departure platforms.

Brunel's first thoughts for Paddington, as committed to his sketchbook, were for a grand central concourse with platforms and tracks on either side: one may presume that one side would have been for departures, the other for arrivals. As we have seen, the directors forced a change of plan. Instead, the middle of the train shed was occupied by a broad trackbed, running through to a group of turntables behind the hotel. The departure side was to the south, the arrival side to the north (Fig 5.1). It was normal for these to be clearly differentiated, but nevertheless unusual for them to be so completely

Figure 5.1
Paddington Station, looking north from the departure side through Brunel's transepts to the arrival side.
[© English Heritage K010978]

separate as Brunel made them at Paddington: departing and arriving passengers were to be kept apart, as in a modern airport. The two sides were linked by a covered footbridge at the east or 'town end', which also connected them to the hotel. Brunel also provided a subway spanning the middle of the station, intended only for staff, though this was later made available to the public as a means of reaching the central platforms. In the middle, beneath the central trackbed, was a sizeable underground room. Its original purpose is not clear: it may have been a 'lamp room' for supplying oil lamps for the trains, but Brunel could have intended it to house hydraulic gear for operating machinery in the station, as discussed later.

Indeed, reconstructing how Brunel intended the new station to work is not easy as he left no general description of this, and there are no surviving drawings on which all the functions are marked. A certain amount can be inferred from various plans. There is also a useful series of articles on 'The Development of Paddington Passenger Station' in the *GWR Mazagine* for 1916 by E C Matthews, a company employee with an encyclopaedic knowledge of the station, who had learnt much from old GWR hands. Even he did not fully understand Brunel's design.[2]

Passengers arriving at Paddington by carriage or cab or on foot would go down the ramp from Eastbourne Terrace to the main entrance loggia, leading onto the departure platform. Just to the right was the main booking hall. In 1845 the GWR adopted the 'Edmondson ticket' system, named for its inventor Thomas Edmondson, in which cardboard tickets were pre-printed and franked with the date on the day of travel (Fig 5.2). The tickets were printed by the million in the GWR's own presses at Paddington: they had monetary value and so had to be kept securely. It was a complex system as colour-coded first-, second-, and third-class tickets had to be printed for all destinations: in the 1860s the GWR was issuing 12 different kinds of ticket to every station on its network.[3] To the right of the booking hall there was the first-class waiting room, with its separate ladies' room and cloakrooms. Left of the loggia was the second-class waiting room and beyond that the royal waiting room and a refreshment room. There was, of course, no third-class waiting room.

The new station was provided with 10 broad-gauge tracks. These are numbered on the accompanying plan from the south (departure) side, going north (*see* Fig 3.9). There was the departure platform, then tracks 1 and 2, then an island platform, and then tracks 3 to 9 under the central span. Track 9 faced another island platform, then came track 10 and the main arrival platform. The island platforms would have allowed two trains to be prepared and loaded (or unloaded) simultaneously. Matthews thought they were also used as 'lamp platforms' being linked via the subway to the lamp room, beneath the central trackbed. In the days when trains were lit by oil lamps, servicing these was a 'heavy and important' business.[4] To provide access to these island platforms, Brunel and Armstrong designed hydraulically powered moving bridges, stored under the departure platform, to be rolled out across the tracks when needed (Fig 5.3). Rather oddly, the plans do not seem to show any such bridges for the island platform on the arrival side.[5]

Brunel was fascinated by the hydraulic machinery being developed by the celebrated inventor and industrialist Sir William Armstrong in his works at Elswick on the Tyne. He collaborated with Armstrong in developing a whole series of machines for Paddington, as if testing how far the station's operations could be mechanised. The bridges and traversers were all to be hydraulically powered, as were the many turntables, sector tables, cranes and lifts around the station (Fig 5.4). All were designed and manufactured by Sir William Armstrong & Company at Elswick.[6] Brunel's office diaries for 1851 record frequent meetings with Sir William Armstrong, and in 1854 Armstrong wrote:

I have also applied it [hydraulic power] extensively to railway purposes, chiefly under the direction of Mr Brunel ... Most of these applications are well exemplified at the new station of the Great Western Railway Company in London, where the loading and unloading of trucks, the hoisting into warehouses, the lifting of loaded trucks from one level to another, the moving of turntables and the hauling of trucks and traversing machines, are all performed, or about to be so, by means of hydraulic pressure supplied by one central steam engine with connected accumulators.[7]

None of this machinery now exists. There is enough documentary evidence to show that Brunel and Armstrong were attempting something very ambitious here, but it is difficult to reconstruct in detail, and a number of aspects remain puzzling. Firstly, it is not clear how the hydraulic machinery was actually

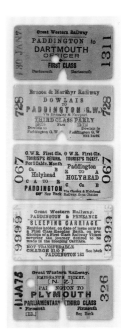

Figure 5.2
In the 19th century the GWR issued pre-printed Edmondson tickets for all its scheduled and special services.
[STEAM – Museum of the Great Western Railway, Swindon]

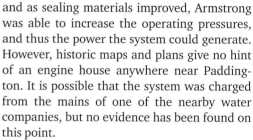

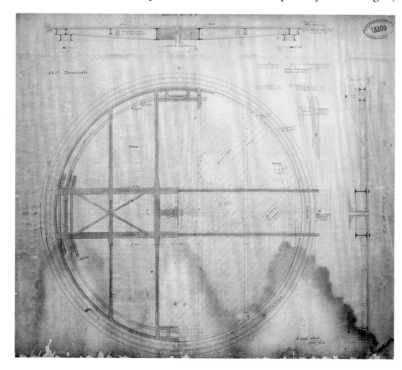

powered. Normally, a system like this would rely on a hydraulic accumulator. The accumulator, invented by Armstrong himself, was a tall pressure vessel filled with water, put under pressure by a massive weight on top of it: by turning a valve the pressurised water was released through pipes to turn or push a piece of machinery, as the weight descended. Steam power was needed to keep the system charged,

and as sealing materials improved, Armstrong was able to increase the operating pressures, and thus the power the system could generate. However, historic maps and plans give no hint of an engine house anywhere near Paddington. It is possible that the system was charged from the mains of one of the nearby water companies, but no evidence has been found on this point.

The other problem relates to the station's transepts, and some features housed in them called 'descending platforms'. Brunel had incorporated at least one large, probably manually operated traverser, in the temporary station at Paddington (*see* chapter 2). There has been a persistent story that Brunel designed the transepts of Paddington Station so as to house traversers, or at any rate to allow for them being installed, across its whole width. The origin of this seems to be his son's *Life of Isambard Kingdom Brunel*, which refers to it being crossed by 'two transepts 50ft [15m] wide, which gave space for large traversing frames'. On the face of it, this would offer a plausible rationale for the transepts: the regular column spacings in the train shed are around 30ft (9m) and in 1852 the GWR introduced new 38ft (11.6m) carriages known as 'Long Charlies'.

There is no other evidence for this, however, and various documents in Network Rail's archives paint a different picture. A foundation drawing, undated but probably from early 1851, shows foundations for two traversers, but not in the transepts. They were around 26ft (7.9m) wide (which was thus the maximum wheelbase length of a vehicle that could be carried on them).[8] Furthermore, they only spanned or linked lines 1 and 2, so they did not need to cross a row of columns.[9] Other layout drawings, datable to 1851, seem to confirm that traversers were only planned for these two positions on the departure side.[10] There are numerous drawings for the traversers and the hydraulic rams to work them, all by Sir William Armstrong & Company of various dates c 1852–4 (Fig 5.5).[11] Finally, there is photographic proof that traversers were built in these positions, one of which remained *in situ* into the 20th century (Fig 5.6).

So far, this seems clear enough: Brunel wanted traversers to communicate between the two departure lines, to take vehicles with a maximum wheelbase of 26ft (7.9m). There would be no need for the transepts on this account. However, other drawings show that the transepts were designed to house 'descending platforms'. Sections of the island platforms, measuring 44ft (13.4m) by 24ft (7.3m), were to be constructed so they could be hydraulically lowered on iron frames to about track level.[12] Another plan has the descending platforms roughly sketched in all four positions (ie both island platforms, in each transept), and labelled '46 feet'. Clearly, they would not have fitted between the standard 30ft (9.1m)-column spacings, but what on earth were they for? The drawings give no clue as to the

purpose of these things. Their decks or surfaces were of plain boarding, with just one feature, a narrow groove or channel in the middle, about 2ft (0.6m) deep and running in the transverse direction (at 90 degrees to the train tracks). It seems impossible that rolling stock could have moved across them – that would have required a traverser, a quite different piece of equipment.

It is not clear if these strange devices were ever installed: nevertheless, this is the only piece of evidence to 'explain' Paddington's transepts, a vital feature of its design. If they were built, they were quickly taken out again and seem to have been completely forgotten. By the time Isambard Brunel Junior wrote his father's life, the story was of the transepts being intended to house a glorified system of traversers. E C Matthews, a well-informed member of the GWR's staff, recorded this story in around 1913, noting that when the platforms were rebuilt in the 1880s 'a mass of girder-work and piping was discovered for which there was no apparent use, and in the writer's opinion this was intended to have formed a part of the traversers, the girder-work for carrying them, and the pipes for operating'.[13] In fact, the girder-work and piping would seem to have been for the mysterious descending platforms, but whether they were installed, and what they were for, we have no idea.

In the 1850s a typical GWR express train would be composed of an engine and tender, a guard's van, and six to eight carriages, divided between the first and second classes.

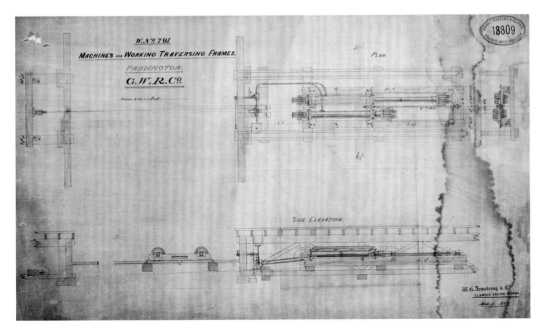

Figure 5.5
'Machines for Working Traversing Frames', a design by Sir William Armstrong & Company, 1852.
[Network Rail/EH AA031073]

The second-class carriages would usually come after the engine with the first-class carriages at the back, corresponding to the positions of their respective waiting rooms as they opened onto the departure platform. Frith's painting (*see* Fig 4.2) has a train with third-class carriages at the front, but the GWR continued into the 1860s to carry most of its third-class passengers on separate trains. Until 1875 luggage was commonly stowed in metal racks on the carriage roofs, covered with tarpaulins, again as shown in Frith's painting.

Wealthy families travelling between London and the West Country would often want to take their horses and carriage with them. In the GWR's early years many richer passengers preferred to travel in their own carriages

strapped to flat carriage trucks, an early Victorian equivalent to the Channel Tunnel's 'Le Shuttle' service. For this, the company tended to charge about a shilling less per person than for standard first-class travel. It is not clear how long this particular service endured, but the handling of private carriages and horses long remained a lucrative part of the GWR's business: in 1849 the company owned no fewer than 224 carriage trucks and they were kept at all major stations (*see* Fig 3.1). The service only needed to be booked the day before travel: horses and carriage had to be at the station 15 minutes before the scheduled departure time. In 1841 the company charged 48s to take a two-wheeled vehicle, or 58s for a four-wheeled vehicle, from London to Bristol. They would take a horse from London to Bristol for 48s, a pair for 73s. The GWR's care for its smarter 'county' passengers was also manifested in its 'hunting services': for example, in 1865 a huntsman could take himself, his groom and a pair of horses to Oxford, Woodstock or Banbury for the season and back, for £18.[14]

Brunel made generous provision for the horse-and-carriage business in the layout of the new station. A wealthy family on their way to the West Country could be driven to Paddington and dropped off at the entrance to the first-class waiting room, while (presumably) a servant collected their tickets. Their coachman would take their carriage to the 'horse arch', at the eastern ('town') end of the cab road. Here, Brunel had designed a loading bay where four tracks ran up to the platform edge, linked to the main tracks by a set of small turntables. This feature formed part of his first sketch for the station layout, so it was clearly regarded as very important. The horses would be led into horseboxes, the carriage run up a ramp and strapped to a carriage truck, and these together would be turned on the turntables, before being manhandled along the tracks and hitched to the back of a train on the departure platform for the journey west. A very similar loading bay, again with four 'docking' positions, was provided at the outer end of the arrival side, presumably for horses and carriages making the journey up to town.[15]

This reconstruction is rather tentative: these turntables had been taken out by 1867, so in the event they were soon obsolete.[16] E C Matthews thought the area may have been designed for passengers who intended to travel in their own carriages on the train and that the justification for it died out with this practice. The GWR would not allow horses and carriages on its express services and it may be that by 1867 its richer passengers preferred to travel by express and send their horses and carriages on ahead by night trains. Even so they must have been loaded somewhere, and it is not clear where this happened in the later 19th century.

A curious detail of the plan was a single track of a narrower, 4ft (1.2m) gauge, just beyond the Horse-Arch turntables, running from the back of the hotel to the end of the departure platform. According to Matthews, this carried a special luggage trolley, presumably to take hotel visitors' baggage to be loaded onto departing trains.[17]

The horse arch could also have been used for mailbags or parcel deliveries (which were always handled by the passenger station, not the goods station). Brunel provided the departure side with another loading bay, however, this one at the 'country end' and actually outside the train shed. This too had a wide vehicle entrance leading onto the platform and four lines running up at right angles to the platform edge, in this instance linked to a separate, dedicated line of track. By 1876 this area was known as the 'milk platform', though it would seem to represent a lot of investment merely for handling milk churns.

The most important point, perhaps, is that the layout was very flexible. The horse arch was designed to allow horses, vehicles, sacks or parcels to be loaded conveniently into vans or trucks and hitched to the back of a passenger train on the main departure platform. The milk platform, outside the shed, could have been used to get churns, horses, vehicles, sacks and parcels onto dedicated services, or shunted onto the backs of departing passenger trains.

The arrival side, on the north flank of the station, was more simply planned and had no really substantial buildings (Fig 5.7). The three spans of the main roof were flanked by a lower aisle covered with flat trusses and glazed Paxton roofing: a long, solid brick wall closed the station on its north side. There were two arrival lines, 9 and 10, on either side of the island platform, so the station could cope with two incoming trains at once. Three relatively small timber-framed buildings stood on the arrival platform: reading from the hotel end, they were the main waiting room, a cloakroom and left luggage office, and the royal arrival side waiting room.[18] Beyond them, under the Paxton roof, was a broad roadway, allowing vehicles to come right down into the station.

At the outer end of the arrival side was a loading (or rather, unloading) bay, like the two on the departure side. Like them, it had four docking positions, so that wagons or trucks could be detached from the back of a train, turned on small turntables and run end-on up to the platform edge. This, too, figures in Brunel's very first sketch plans. It presumably served for the unloading of horses and carriages and may have been for milk deliveries as well. It seems unlikely that mail and parcels were dealt with here, as the main parcels office, a long framed building, was 200yd (183m) away under the far end of the arrival platform.[19] If we assume that parcels were often carried in the guard's van and that in the early days this was usually right

behind the locomotive (as in Frith's painting), then an incoming train would almost deliver them to the door.

One may picture a train arriving from Bristol soon after the opening of the arrival side. It would halt briefly outside the shed, to allow the carriage trucks and horseboxes to be detached. The train would steam into the shed, coming to a standstill parallel with the waiting rooms, cabs and carriages would be waiting in the roadway nearby to greet the arrivals. The locomotive would be detached from the carriages and run forward into the V-shaped 'sector table', diverted onto line 9, and reversed out to the engine shed at Westbourne Park.[20] Porters would place wheeled ladders against the carriages, unload the luggage from the roof racks and put it in labelled bins on the platform, as provided for in the GWR's rule book:

All articles of London luggage not taken charge of by the passengers themselves are to have red labels pasted on showing the initial letter of the owner's surname, and the porters at Paddington are, on the arrival of the trains, to take care that the luggage thus labelled is placed in the proper bins or divisions of the barrier on the platforms, and delivered to the proper owners.

There was an old Paddington story of a porter who told the Reverend Mr L—, enquiring after his luggage, to go to 'L', and being reported for insolence. One hopes the porter was not sacked: he could quite easily have been.[21]

As we have seen, beneath the wide central span there was a broad trackbed with seven lines, only the outer of which (lines 3 and 9) faced onto platforms. This had formed a feature of Brunel's first sketches for the station and had parallels at most other Victorian termini: these platformless tracks seem to have been used for marshalling new passenger trains, or for holding carriages and wagons in readiness. E C Matthews made the interesting observation that Brunel might have intended the central span to be occupied by the locomotive department: this would represent something of a parallel to his original design for Bristol Temple Meads, but I have not found any other evidence for it.[22] The station certainly was not fitted out for this purpose: the lines ran through to turntables, probably 22ft (6.7m) in diameter. These could have coped with the GWR's old six-wheel coaches, which were relatively light and whose wheelbases were only 18ft 4in. (5.6m) long. They could not have coped with the new 38ft (11.6m) eight-wheel carriages, the 'Long Charlies', which in 1852 were being introduced for the Birmingham service.[23] Nor would they have been big or heavy enough to handle locomotives. So the group of turntables was presumably for handling short carriages and mail vans, which could thus be moved from any one line to another, and regrouped to form new trains. As it turned out, the turntables seem to have been rendered obsolete and were taken out again around 1867.

The evidence for all this is incomplete and, in the case of the descending platforms, positively mysterious. What emerges is the impression that Brunel was thinking in terms of an 'ideal station', in which muscle power would be replaced by steam power, and rolling stock could be moved around with the greatest freedom possible. It may be that this vision of a mechanised station, like some of his other projects, proved to be over ambitious. In the long run, the station's operations evolved in a quite different way, and the real value of the central span and of the Lawn area to the GWR was simply as growth room.

One passenger mattered above anyone else to the Great Western Railway, and that was Queen Victoria. By good fortune the journey between the queen's two principal residences, Buckingham Palace and Windsor Castle, fell within their territory. The queen was persuaded to make her first train journey, from Slough to Paddington, on 13 June 1842. She professed herself to be 'quite charmed' with rail travel, and thereafter her journeys between the two palaces were almost invariably by the Great Western.[24] Soon, the London & South Western were pushing their line from Staines towards Windsor, and the GWR had to fight back, opening their short branch to Windsor Central Station in October 1849, just two months before the enemy opened Windsor and Eton Riverside. The GWR went to great lengths to maintain their royal patronage: as we have seen, the new Paddington was provided with royal waiting rooms, the only one of the London termini to be thus equipped (see Fig 7.21), and their station at Windsor was disproportionately large and splendid for so small a town.[25]

These royal journeys from Windsor to Paddington and back may not have been long but they required precise organisation and elaborate protocols were worked out. Notification was sent from the palace to the general manager well in advance. The GWR would send out

Figure 5.8
Platform 1 decorated for a visit by the Shah of Persia in 1889.
[Reproduced by permission of English Heritage Howarth-Loomes Collection BB82/12079]

a printed notice to all the stationmasters on the route, who had to indicate receipt of it by telegraph, twice. The stationmasters had then to notify every signal box on the route in person. On the day of the journey, one of the lines would be cleared of all other traffic: flagmen covered the whole distance within sight and hailing distance of each other. A pilot engine ran in front of the royal train and either the general manager or the superintendent of the line would be travelling on it.[26] Paddington became a major venue for various ceremonial events – for example visits by foreign royalty and dignitaries (Fig 5.8) – and, being on the official route to Windsor, it was used as a staging post on such great state occasions as the funerals of Queen Victoria, Edward VII, George V and George VI (Figs 5.9 and 5.10).

For the first 40 years or more of its existence, the GWR was about the most class conscious and exclusive of the major British railway companies. The directors saw little profit and no benefit in carrying working people for tiny fares, and their first timetables only offered first- and second-class services (Fig 5.11). J C Bourne, in his celebratory account of the railway of 1846, remarked on the high proportion of first-class traffic, and the number of passengers travelling with their private carriages. At Slough 'the greater part of the traffic was of a very high description'.[27]

In its first two years, the GWR formally offered no third-class service at all. They ran goods trains hiring out wagons to carriers, and from 1838–40 they allowed a number of carriers to convey 'goods passengers' in open

Figure 5.10
King George V's funeral: the king's coffin is about to be placed on a train pulled by Locomotive 4082 Windsor Castle, *28 January 1936. [© National Railway Museum/Science and Society Picture Library]*

wagons fitted with wooden seats, absolutely unprotected from the weather: The seats were 18in. (0.45m) off the floor, the sides and ends only 2ft (0.6m) high. From 1841 the company officially carried third-class passengers, still in open trucks. However, the GWR would not have them slowing down passengers of quality and the third-class services left Paddington at 4.30 am or 9.30 pm.[28]

To modern eyes this looks callous. It may have seemed less so to contemporaries, accustomed as they were to the rigours of long-distance journeys on horseback or on foot, or being jolted for mile after mile in (or worse, on the outside of) the mail coach. Brunel had astringent views on the subject, telling one of his assistants that he 'should travel outside (the coach) – inside is, by day – in England – only fit for women and invalids'.[29]

Early Victorian Britons did not expect travel to be comfortable. Nor was it cheap. In 1841 one would pay 30s to travel from London to Bristol in first class (in a comfortable, fully glazed carriage with padded seats), 21s in second class (in an unglazed carriage with seats 15in. (0.38m) wide), or 12s 6d in the third class (in an open truck on a different, slower train). All this was in an age when an average industrial workman's weekly wages might be £1 or less.[30]

The GWR were not alone in this excessively class-conscious approach, but they were widely regarded as the worst offenders. *The Times*, in a leader of December 1842, said: 'The manner in which the GWR Company treats this class of passengers is described as worse than any other.' The service was dangerous too, a point tragically demonstrated by the Sonning accident of Christmas Eve 1841. A 4.30 am goods train, pulling two wagons of third-class passengers and 17 goods wagons ran into a landslip in the cutting in the dark. The locomotive buried itself in the landslide. The passenger

Figure 5.11
An early GWR timetable,
1839.
[Great Western Trust
Collection]

these would resemble cattle trucks furnished with narrow wooden benches, and having no windows, only ventilator grilles high up. This was the golden age of liberal *laissez faire* economics, and it is a measure of the strength of public feeling that the Government felt obliged to legislate. Gladstone, nevertheless, did not feel that he could dictate the hours of these 'parliamentary trains'. As a result, until 1860 the GWR's normal services still only carried first- and second-class passengers, and their third-class 'parliamentary' service left Paddington at 7 am: the company's servants referred to it disdainfully as the 'Plymouth Cheap'. Alternatively, there was the 'Night Goods' passenger service which departed at 9.30 pm. A third-class passenger leaving Paddington at 7 am would get to Bristol in nine hours with an hour-and-a-half stop at Swindon: first- and second-class passengers would make the journey in a third to a half of that time.[32]

The GWR's class-conscious culture was gradually diluted by competition and social change. In 1875 the Midland Railway astonished the industry (and the public) by abolishing the second class altogether, while upgrading third-class carriages with such second-class comforts as upholstery and heating. The GWR and the London & North Western, the 'old guard' of the industry, were offended by the Midland's vulgar competitiveness, and said so.[33] The GWR were nevertheless finding that their initial vision of the railway as a smart mode of transport for smart people was fundamentally flawed and that the future of their business lay in providing services for the masses. In 1872 the general manager reported that in the past year first- and second-class receipts had dipped, while third-class takings had risen by £268,000.[34] From 1882, third-class carriages were provided on all trains except for the two fastest express services, the *Flying Dutchman* and the *Zulu*. They, at last, took third-class passengers from 1890.[35]

By the 1890s the volume of third-class traffic had vastly outstripped the other two: in 1893 the GWR carried 1.5 million first-class, 4.5 million second-class, and 55.5 million third-class passengers.[36] From being a despised low-margin service which they almost had to be forced to provide at all, the third class had become the bulk of their passenger business. Like the Midland Railway, the GWR found that it was the second class which was losing its *raison d'être*. The company's general manager, interviewed in 1894, said 'the traffic is a

wagons were at the front behind the engine, and as a result the laden goods wagons crushed them against the locomotive tender. Of the 38 passengers, 8 were killed outright and 17 seriously injured. In response, the company raised the sides of its passenger wagons.[31]

The railway companies aroused such public anger that in 1844 W E Gladstone, then President of the Board of Trade, passed the Railway Regulation Act which obliged all the passenger railways in Britain to provide at least one 'workmen's train' a day on every line, stopping at all stations and charging no more than a penny a mile. The Act also obliged the companies to protect their passengers from the weather: the GWR had to provide third-class carriages instead of open wagons, but to modern eyes diminishing one, and the day will come when it will extinguish itself'.[37] In the same year, the company abolished the differential between their full third-class and 'parliamentary' fares, at the lower 'penny-a-mile' rate. The second class was finally abolished in October 1910.[38]

The carriages very gradually became more comfortable. For the first generation even first-class carriages were unheated. In 1856, first-class passengers were supplied with metal foot-warmers filled with hot water or acetate of soda: these were extended to the second class in 1870 and to the third class in 1873.[39] By 1870 carriages were, rather tellingly, divided into 'best express', 'second best', 'ordinary', and 'excursion': the excursion carriages were 'never to be used in regular trains except in case of absolute necessity'.[40] The company continued to build spacious 10ft ¾in. (3.07m)-wide wide-bodied coaches until 1882: after that point, they were built with bodies of 8ft ¾in. (2.47m), designed to be easily converted to the standard gauge. The GWR introduced the first true 'corridor train' in Britain in 1892, on its services to Birkenhead.[41] Although the company offered refreshments on its platforms (Fig 5.12), it was less quick to provide dining cars, only putting them on in 1896, 14 years behind the Midland Railway.[42]

The company had been rather quicker to see the possibilities of the tourist business. The idea of certain places being holiday destinations, spa towns and seaside resorts, was already well established and these places were greatly helped by the railways (Fig 5.13). The GWR's

Figure 5.12
Platform catering: a GWR tea trolley at Paddington, early 20th century.
[STEAM – Museum of the Great Western Railway, Swindon]

Figure 5.13
Holiday crowds for the Torbay Limited, 21 August 1926.
[© National Railway Museum/Science and Society Picture Library]

timetable for 1865 listed tourist services for the south-western coastal resorts in its own heartland, but also advertised services provided jointly with other companies to Yorkshire, Buxton, the Lake District, North Wales, Scotland, Ireland, Boulogne and Calais.

This was not exactly the start of a boom in mass tourism: given the price of the services they were effectively for the upper and upper middle classes. In 1865 a return ticket from London to Penzance cost £3 18s in first class and £2 16s in second class, while a tourist ticket to the Dublin International Exhibition via Holyhead was £3 or £2 7s 6d.[43] Until the First World War, a sum of this order would have represented the weekly wages of one of the GWR's clerks. Nevertheless a trend had been established, and the services grew steadily. In 1852 Measom's *Illustrated Guide to the Great Western Railway* was published, with a second edition coming out in 1861, followed by Cassell & Co's new *Official Guide* of 1884. Essentially tourist guides to the historic and natural beauties of the GWR and its allied companies' territory, they provide clues to the nature of this part of the GWR's business and clientele in the shape of numerous advertisements, for example for hotels in Ilfracombe and Torquay, various private schools, porcelain, gentlemen's outfitters and bathing costumes ('he won't spoil his things, Auntie, they are made of Spearman's Royal Devonshire Serge, like yours and mine!').

Holidays in the West Country may still have been the preserve of the rich, but the GWR was also developing a busy traffic in day trips up the Thames valley for the less well off. In January 1863 the Lord's Day Observance Society wrote to the directors, enclosing a memorial signed by the archbishops and bishops of England, 'urging the discontinuance of Sunday Excursion trains'. The directors robustly refused, maintaining that the trains referred to 'are conducive to the good conduct as well as the innocent enjoyment and recreation of the poorer classes for whom they are principally promoted, and without leading in any way to the desecration or improper observance of the Sabbath … '.[44]

The GWR, like other companies, was not averse to promoting prize fights in order to drum up business: in 1863, when the company put on a special train for Mace versus Gove at Wootton Bassett, Wiltshire, there was a serious riot at Paddington. Victorian respectability, however, was catching up with this kind of thing, and a Regulation Act of 1868 forbade the companies to put on 'specials' for prize fights.[45]

In 1877 the GWR was advertising return tickets to Windsor for 2s 6d or to Henley for 3s 6d: these were affordable for almost anyone

Figure 5.14
A GWR wagon and four outside the Mint stables, probably 1920s.
[Great Western Trust Collection]

in employment.[46] In 1881, after an especially busy Bank Holiday weekend, the company decided to introduce an extra platform at Paddington especially for 'excursion traffic'.[47] The company's general manager, interviewed in 1894, talked about how heavy the tourist traffic had become: the GWR had joined forces with the boat operators Messrs Salter, to sell combined tickets for days out on the river at Goring and Henley. It had introduced 'Friday night specials' to the West Country, returning the following Monday, or Monday week. In 1880 the GWR had 2,147 miles (3,455km) of track, and appears to have carried 44,500,000 passengers a total of 13,202,399 train miles. By 1893 the track mileage had risen to 2,494 (4,013km), but the company carried 61,250,000 passengers a total of 18,817,118 train miles.[48] Rail travel was becoming more accessible for the Pooters as well as the Forsytes.

Goods traffic had also come a long way since the 1840s, when the depot usually managed two incoming and two outgoing trains most days. After the opening of the Chepstow bridge allowing through-trains to South Wales in 1852–3, a huge traffic in coal developed and Brunel laid out the 'high-level goods yard' alongside the canal basin expressly for this. The GWR had probably the most extensive trade in foodstuffs of any of the major companies and began to develop special facilities for handling it. In 1863 the company opened a great new depot below the newly built Smithfield Market in partnership with the Metropolitan Railway.[49]

In the 1850s and 1860s the GWR developed their goods, parcels and milk delivery services in London, depending on carriers, hired horses and rented stables for the purpose. By the 1870s this no longer seemed adequate. The other large companies were moving towards having their own establishments and in 1875 the company resolved to build its own stables at Paddington and at Smithfield and Crutched Friars in the City of London. The most important was always the Mint stables, just north of Paddington, named for the Mint tavern on London Street. The first stage, with room for 288 horses, was built between 1875 and 1877 at a cost of £25,300. In December 1877 the company decided to start buying its own horses.[50] By 1894 they had 1,839, compared to the Midland Railway's 4,346. The stables continued in use surprisingly late, until after the Second World War, and indeed, until after nationalisation (Fig 5.14; see also Figs 8.23, 8.24, 8.25 and 8.26).

It is very difficult to estimate the number of Great Western Railway staff at Paddington in the later 19th century. Certainly there were many hundreds, well up from the 269 counted there in 1843. A number of staff lists have been found for 1873. A 'room allocation' document lists 76 'head office' staff in the chairman's office, the accounts department, the general manager's and engineer's departments and other central functions.[51] Another 41 managers and clerks ran the station operations[52] and a further 51 ran the goods station, giving a total of 168 'white collar' staff.[53] No lists have been found, however, for the great numbers of porters, general labourers, carters and ostlers who were employed in the station, parcels office, goods depot and stables. Nor are there any lists of the train and track staff, the enginemen, inspectors, signalmen and guards who were based there. The grand total in 1873 would have been unlikely to be less than 500, perhaps nearer 1,000.

In later years, the GWR had a well-deserved reputation as a benevolent employer, but for its first 50 years the company treated its staff harshly, like many other Victorian enterprises. The harshness was tempered with incentives. The company was meritocratic, rewarding zealous service with promotion. It had pay scales with maximum salaries attached to each post to reward staff for long service.[54] Enginemen who gave good service without accident or cause for complaint for a year were given a standard bonus of £10, while efficient switchmen would receive a bonus of £5.[55] The company, however, maintained ferocious standards of industrial discipline. Until the 1860s the board handled serious disciplinary matters themselves: the accused were summoned to Paddington to stand at one end of the directors' polished mahogany table, an ordeal known to the staff as 'going to see the Picture' (Fig 5.15 and see Fig 4.1). The board minutes record great numbers of cases dealt with at alternate (that is, fortnightly) meetings, usually with brutal swiftness. A few typical examples from the early months of 1855 should give the modern reader a hint of what Victorian staff management techniques were like:

11 January 1855
John Peckham and George Hobbs, switchmen at Chippenham, were called in, and examined on a charge of being in the Refreshment Room at the station instead of attending to their duty. They expressed much regret at the cause of complaint, and

their previous conduct having been exemplary they were only fined 20 shillings each with a caution for the future.[56]

Mr Graham reported that Henry Griffiths, a Policeman at Bath, had been brought before the magistrates on a charge of stealing a lady's jacket, but sufficient proof of his guilt had not been given to convict him. As there is no doubt, however, of his having stolen the jacket, he was ordered to be dismissed the service.[57]

25 January 1855

Jones, a porter at the Gabowen Station, was reported for being intoxicated, and the charge being proved, he was dismissed.[58]

Huckle, a porter at Didcot, was called and examined on a charge of attempting to give a light to a passenger for the purpose of lighting a pipe, which being proved he was severely reprimanded and fined 20 shillings for not telling the truth when charged with the offence. He was informed that had it not been for his very long and good service during 16 years he would have been dismissed.[59]

8 March 1855

Received a letter from Mr Fryer intimating that Mr Cooper, a Station Clerk at Harbury, had died in consequence of a Locomotive Engine having passed over his foot while in the discharge of his duty, and had requested that the Directors would allow him a month's salary, to be given to his Mother and Sister, who were in very indigent circumstances, after his decease, when the sum of £10 was ordered to be given to them. [60]

These are five examples out of hundreds: the shadow of the workhouse seems to loom over these dismal proceedings.

In the 1860s the GWR was in a very difficult financial position, and this seems to have driven the company towards harsher policies. In 1863 the general manager commissioned a report on staff and working expenses at all of its stations. The document is a monument of absolute utilitarianism: no considerations of loyalty to long-serving staff were to be allowed to stand in the way of economy. The big idea was to replace experienced employees with children. The second-class booking office at Paddington could handle the third-class business as well, so three clerks at £70 per annum each could be replaced by three boys at £20 per annum. Four porters could be replaced by six young men on trial at 12 shillings a week. Several clerks at Paddington goods depot had been 'many years in the Company's service' and thus advanced to high salary levels: nine of them in the 'abstract office', receiving a total of

Figure 5.15
The Great Western Railway's boardroom, c 1900, dominated by a great map of the network. A bust of Brunel is just recognisable on the mantelpiece. The full-length portrait of the great chairman Charles Russell, known as 'the Picture', must have been at the other end of the room.
[© National Railway Museum/Science and Society Picture Library]

£694, could be sacked and replaced with junior clerks, saving £150, and so on. One could imagine Dickens having a field day with this depressing document.[61] It is not clear whether the ideas were implemented, though in view of the company's plight they may well have been.[62]

The GWR's staff were not, however, without cultural resources. In 1843 a Mechanics Institute with a library was founded for them at Swindon 'for the purpose of disseminating useful knowledge and encouraging rational amusement among all classes of people' employed by the company.[63] This does not, however, seem to have been replicated at Paddington. By 1849, though, the company had established a station library at Paddington, with over 1,000 volumes, principally novels, for the use of the travelling public, who had free access to the books while waiting for trains for 1d, or they could hire a volume and turn it in at their destination for slightly more.[64] The GWR Literary Society, established there in 1852, seemed in its early years to embody the class divisions within the workforce, being reserved for the management and the salaried, office-based employees. It started with a library of 730 volumes and a generous budget provided by the company, and it charged subscriptions of 2s 6d a quarter.[65] The directors supported

the new society, and the chairman, the Earl of Shelburne, became its first president.

This does much to explain a remarkable event, a 'Conversazione of the Great Western Railway Literary Society', held at Paddington on 7 to 9 December 1859 and described by *The Illustrated London News* (Fig 5.16). The company's large meeting rooms were decorated with 'beautiful and rare specimens of pictures, sculpture, bronzes, books and other works of art' lent by the chairman, directors, and officers. The centrepiece was an elaborate bedstead presented to Queen Victoria by the Maharajah of Kashmir. Autograph letters of Charles I and ancient jars and vases excavated by Robert Stephenson in Alexandria were on display. The rooms were hung with works by eminent contemporary painters including Samuel Reynolds, David Wilkie and Clarkson Stanfield. Several companies lent pieces of decorative artwork: ceramics from the Copeland company, bronzes from the Coalbrookdale company, decorative papier-mâché, gold and silver ware. A central window was 'arranged as a memorial to the genius of Brunel and Stephenson', both of whom had recently died. In this elaborate setting the society held concerts and poetry readings for three successive evenings, at which 'several amateur friends gave their services, including a reading from the "Ingoldsby

Figure 5.16
The 'Conversazione of the Great Western Railway Literary Society', Paddington (from The Illustrated London News, *17 December 1859).*
[City of Westminster Archives Centre]

Legends" by the honorary secretary of the Society, Mr Kinnaird'.[66]

The conversazione seems to have been a one-off, but the society prospered. There were lectures, a dining club was founded in 1860, and an 'essay class' in 1861 (members read essays to each other on subjects like 'Tennyson', 'The Qualities Necessary to Success in Life', and 'The General Outlines of the Science of Ethnology'). In 1862–3 the *Great Western Railway Magazine and Temperance Union Record* was established for the first time. It recorded the Literary Society's growth: membership up from 403 to 416, the library collection up to 4,253 books. A member of the society, writing anonymously to the magazine, urged that its membership and reading room be opened to the 'second class' of staff – to inspectors, guards, drivers, mechanics and the like.[67]

The *GWR Magazine* was re-established in 1889, soon dropping the 'Temperance Union' from its title. It seems to have been formally adopted by the company and from then on appeared monthly, right up until the end of the GWR in 1947. The magazine exudes an intense corporate pride. Much space is devoted to engineering projects and rolling stock, but there are also substantial articles about the company's history, its stations and its art collection. There are regular profiles of staff in the various departments.

Every issue carries accounts of staff dinners: the 'down loading section' of the goods department, for example (dinner followed by toasts, harmony, recitations, and the national anthem). Regular reports appear for any

number of societies: the Pelican or Swimming Club, the ambulance classes, the Boating Club (annual regatta at Staines), the Football Club (founded in 1898, with a ground at Acton), the Paddington Madrigal Society and the Chess Club among others. In February 1898 the magazine reported the second concert of the 40-strong GWR Orchestra at the Ladbroke Hall: a programme of Gounod, Sullivan and Suppé, with numerous encores. Regular events were held in aid of the Widows and Orphans Fund, whose famous collecting dog, Tim, became a familiar sight at late Victorian Paddington, greeting arriving trains on behalf of the fund for more than 20 years. GWR employees made up a whole company of the 2nd South Middlesex Volunteer Rifles: space was found in the goods station for a rifle range. In 1896 Messrs Elkington, the silversmiths, presented a shield for their annual shooting competition: it had pride of place in the 'general meeting room' at Paddington.[68] In the pages of the magazine, the company seems a kinder place to work, with a rich social and cultural life for all its employees (Fig 5.17).

In its golden age, *c* 1875–1914, and its silver age, *c* 1923–39, the Great Western Railway had as rich a corporate culture as has ever existed in this country, which endured right up to the Second World War and nationalisation. If one were to take a single episode to typify it, it might be an 'Exhibition of Art at Paddington Station' described by the *GWR Magazine* in 1925 in which 232 paintings, woodcarvings and embroideries, all by the Paddington porters, were exhibited in the 'seat registration office'. Various directors and senior managers showed willing and purchased items at the opening, in aid of the Widows and Orphans Fund. Cheerful events such as this never, of course, give one the whole picture: in the following year the GWR, like its rivals, was seriously affected by the General Strike.

By 1930 the Great Western Railway may have been in difficulties, but it had reached its greatest extent and its highest pitch of sophistication and complexity. In 1929 it had 8,900 miles (14,300km) of permanent way, representing 3,800 miles (6,100km) of route mileage, serving over 1,500 stations. It had over 110,000 employees, 86,000 freight vehicles, 10,000 passenger-train vehicles and 4,000 locomotives. It carried 131,000,000 passengers a year, not including season ticket holders and over 75,000,000 tons (76,203, 518 tonnes) of

Figure 5.17
Members of the Swindon GWR band, practising at Paddington for a competition at the Crystal Palace, 29 September 1934.
[Getty Images]

merchandise and mineral traffic. The company had 2,650 horses, 212 miles (341.2km) of canal, 16 steamships, and owned a larger area of docks than any other organisation in Britain.[69]

Paddington, the hub of this great industrial empire, was a world in itself. In January 1937 Mr F W Green, the assistant divisional superintendent of the line, gave a remarkable lecture on 'The Working of Paddington Passenger Station' to the GWR Lecture and Debating Society. Subsequently printed in the Society's magazine, it provides an unequalled insight into the life of the station in the GWR's silver age.[70]

The station then had 18 platforms, 15 for passengers, 3 for parcels and milk. Of the 15, 11 were around 1,000ft (300m) long for long-distance services: 4 on the departure or down side, 5 on the arrival or up side, 2 for in and out services. The other 4 were the suburban and underground platforms on the north side of the station, about 600ft (180m) long. Four signal boxes (departure side, arrival side, suburban and Westbourne Bridge) regulated the traffic. Every day, 21 well-loaded long-distance trains reached Paddington by 10.30 am. Engines were stabled out at Old Oak Common, or at a subsidiary depot by Ranelagh Bridge: on the previous August Bank Holiday 165 of them left the shed.

Trains were formed out at the carriage sheds, also at Old Oak Common.[71]

Over 2,000,000 passengers bought tickets in a year, at four booking offices (two on the departure side, one on the footbridge leading to the suburban platforms, one 'passimeter office' containing automatic ticket machines at the Bishop's Road entrance) (Fig 5.18). For special events like Newbury races, portable offices could be put up on the Lawn. On 21 July and 1 August 1936, the peak days of the year, 14,646 and 16,238 passengers were booked. They could reserve seats at a separate reservation office on the Lawn: up to 25,000 seats could have been booked in a day, helped by the company's telephone system (Fig 5.19). Diagrams for the next day's trains were put together in the late afternoon and sent to the superintendent at Old Oak Common. The handling of taxis and cars was already a problem: over 6,000 road vehicles could pass through the station in a day, peaking on the departure side between 10 am and 11 am.

It is startling to find that the passenger station handled far more than just passengers. Most of the GWR's market traffic, for example, passed through it. Anything from 50 to 200 tons of flowers, fish, fruit, meat and vegetables

Figure 5.18
The Paddington ticket office, with pre-printed Edmondson tickets, 1920. The system required the production of printed tickets to and from every destination in the GWR network.
[© National Railway Museum/Science and Society Picture Library]

came into the arrival side early each morning: 38 men were needed to handle it and 28 horse vans were on hand to take it away, as well as 15 to 20 lorries. Fruit was also sold in the station (Fig 5.20).Then there was the milk traffic. At one time vast numbers of churns (2,000,000 per annum in 1929) were delivered to the arrival side every day to be manhandled onto vans, with the accompanying 'difficulty and noise' of returning the empties from the down side. This had caused so many complaints from neighbours that the handling of the empty churns had been transferred to the cattle platform at the goods station. The GWR was replacing the 'churn service' with tanks, and a new milk depot at Wood Lane.

The newspaper traffic was on a similar scale: 203 tons (206 tonnes) of newspapers were forwarded onto night trains in the passenger station every night, rising to 394 tons (400 tonnes) on a Saturday. This business employed 106 men on weeknights, 160 on Saturday nights, with 7 tractors for the platform trolleys.

The GWR had long encouraged the growth of special services. More than 2,000 cases of pastries from Messrs Lyons the bakers, for example, were handled each day, while hundreds of tons of poultry passed through the station in the few days before Christmas. The ice-cream traffic was an important departure side business in hot weather: in August 1936 over 350 tons (356 tonnes) of it was sent outwards, in 6,000 containers. Bullion was also sent by rail – more than 20 tons (20.32 tonnes) had been sent in one recent month, presumably late in 1936.

These services were dwarfed by the parcels and mail departments. The parcels depot, just beyond the Bishop's Road Bridge on the departure side, had its own platform 1,000ft (300m) long, a through line on one side, a 600ft (180m) siding on the other. A tunnel connected it to the parcels subway, to platforms 1 to 8. On average, 22,000 parcels were forwarded a day, and 1,400 received for delivery in London, or transfer to other companies. Over 95 motor vans were kept in the parcels depot, in addition to the legion of horses in the Mint stables. Seven four-wheel tractors were kept on the down side, four on the up side, with one spare.

In addition to all this there were the Royal Mail services coming through the Paddington

sorting office on London Street. In an ordinary day it processed 4,500 mailbags and 2,400 bags of parcel post. Down mail went from the Post Office via a conveyor in a tunnel to the south-east corner of the Lawn, onto 'brutes' (wire-sided buggies) and then to waiting trains. Up mailbags, usually arriving on platform 11, were dropped down chutes onto another conveyor that took them to the London Street office for sorting, from where they were sometimes transferred on to Liverpool Street on the Post Office's underground railway.[72] During the Christmas rush the goods department had to help out, setting up

a temporary depot on Alfred Road: over Christmas 1936, 79,389 parcel-post bags were sent in 28 special trains.

All of this teeming activity was handled by 723 station staff and 711 parcels depot staff, not including the extra hands taken on to cope with seasonal business. These, of course, were quite distinct from the 1,000 or so head office staff based at Paddington, and the hundreds more employed in the goods station. Since nationalisation much of this activity has gone, but new services have arrived. Paddington continues to evolve, and it remains one of the busiest passenger stations in Britain.

Figure 5.20
'Eat Empire Fruit': a stall on platform 1, 15 July 1927.
[© National Railway Museum/Science and Society Picture Library]

Decline, Revival and the Future, from 1948

With the GWR abolished, Paddington became the headquarters of the British Railway Board's Western Region. In 1949 a rededication of the war memorial, to commemorate the GWR employees who had fallen in the Second World

Figure 6.1 (right)
The new screen in 'Festival of Britain' style across the platform heads, with the arms of the counties served by British Rail's Western Region, 1950s.
[Network Rail]

Figure 6.2 (below)
Over a thousand West Indian immigrants arrive at Paddington, 9 April 1956.
[Getty Images]

War, also represented a final farewell to the old days. There was no immediate prospect of new investment: the Attlee Government was chronically short of money, and it was not until 1952–3 that the main war damage to Paddington was made good. The main roof was restored, though it seems that the ridge-and-furrow Paxton glazing was removed from 'span one' in the process, and the gap in the Eastbourne Terrace buildings was filled with a simple contemporary structure housing a new booking hall and ticket office in the place of the old one. There was no money for frills and no attempt was made to make them match the style of the original buildings.

The centenary of the station in 1954 was celebrated with the unveiling of a commemorative plaque. One further alteration represented a welcome introduction of 1950s style: a new ticket barrier screen in a 'Festival of Britain' manner was installed in 1958, with panels representing the arms of the counties served by the Western Region (Fig 6.1). Much of the station remained grimy and cluttered, however, even though Paddington was now welcoming a host of new passengers (Fig 6.2) and still loomed large in British popular culture and public awareness. In 1957 Agatha Christie gave the station a walk-on part in her novel *4.50 from Paddington*, and the author Michael Bond gave the famous name a new layer of meaning for generations of children with his Paddington Bear books, the first of which appeared in 1958 (Fig 6.3).

In retrospect, it may have been fortunate that there was not more money available for Paddington (Fig 6.4). In 1952 a Russian-born architect, Sergei Kadleigh, had already produced speculative designs for a vast high-rise development immediately outside the station: 'High Paddington' would have housed 8,000 people in high-rise towers on an 18-acre (7.3ha) podium on the site of its goods yard.

Mercifully, this Corbusian nightmare never stood any chance of being realised.[1] However the 1960s did see the complete demolition and rebuilding of their old rivals the London & North Western Railway's two principal stations, Euston (1960–8) and Birmingham New Street (1964–7). Many other fine Victorian stations were demolished in the 1960s and 1970s, but the demolition of Euston turned out to be a turning point, galvanising public opinion and giving the conservation movement a national cause and national standing for the first time. In the 1970s, under the chairmanship of Sir Peter Parker, British Rail began to re-evaluate its architectural heritage. A wider revival of interest in Victorian architecture was under way and Paddington's outstanding quality

and interest was being recognised: as long ago as 1951 it had been the subject of a ground-breaking article by Henry-Russell Hitchcock in the *Architectural Review*[2] and in 1961 it was listed. In contrast to this at the end of 1965 steam power ceased to be used on British Rail's Western Region, the end of the engineering tradition of Gooch, Churchward and Collett.

In 1968–9 British Rail spent large sums on alterations to Paddington, by their own Western Region architects and engineers together with Seymour Harris & Partners. More of the main shed roof was repaired and re-glazed, involving the removal of the original Paxton roofing in favour of a simpler system. New electrical systems and lighting followed, along with a new ticket office, travel centre and platform

Figure 6.3 (above)
'Please look after this Bear'.
The publication of Michael Bond's first 'Paddington' book in 1958 gave a new meaning to the famous name. A sculpture on the Lawn provides a permanent reminder of this literary connection.
[© English Heritage K030903]

KEY TO NUMBERS

1 Church
2 Apartments
3 Offices
4 Existing Offices
5 Vehicle Ramps
6 Shops
7 Self-Service Supermarket
8 Cinema and Swimming Baths
9 Flower Gardens
10 Green Roof Gardens
11 Central Court : Exhibitions etc.
12 Car Park beneath (1000 Vehicles)
13 Nurseries, Clubrooms etc.
14 Town Hall
15 Children's Playground
16 Primary School
17 Classrooms with Open Terraces
18 Administration
19 Playgrounds
20 Clinic and Isolation Hospital
21 Observation Bar
22 Restaurant and Dancing
23 Meteorological Station
24 Hotel
25 Commercial Television Studios
26 Roof Gardens
27 School recreational grounds
28 Parks

Figure 6.4 (left)
'High Paddington', a scheme for a nightmarish 'dream city' over the tracks and goods station (excerpt from the illustration published in Picture Post, *12 March 1955).*
[Getty Images]

Figure 6.5
The redecorated restaurant,
March 1974: a spirited
exercise in 1970s taste.
[Network Rail]

barriers, and the tracks were curtailed by 40ft (12m) to make a more generous circulating area.[3] The revival of the station may be said to have begun. After this point, an increasing awareness of the station's architectural and historical importance began to have a direct influence on works to it, although some refurbishments continued to reflect contemporary style rather than historical precedents (Fig 6.5).

Another bout of works in 1982–5, handled by the Western Region's architects' and civil engineers' departments, added a new 'tin-shed' depot on the north-eastern side of the station for Red Star parcel services and newspaper handling, together with some refurbishment of the offices.[4] In 1985 British Rail moved its Western Region headquarters to new buildings in Swindon, ending Paddington's long history as the hub of a great network, and the GWR's old offices were refurbished for commercial letting. The station received a form of compensation for this loss of status: British Rail's architects produced the first master plan for managing it better. This produced a shortening of the platforms to increase the passenger concourse, a new ticket office and the first efforts to restore some of Paddington's historic character. The royal waiting room was refurbished for business-class passengers and the clock arch on platform 1 reopened (all in c 1985).[5] This was a prelude to the major restoration of the three Brunel spans between 1985 and 1993, again by British Rail's Western Region civil engineers' department, under George Connell as

project engineer. This was the fullest and most thorough repair of the roof ever to have been undertaken. The ironwork was cleaned right back, with up to 20 layers of paint removed. Serious corrosion was found, especially where the decorative cast-iron plates had been bolted to the arched ribs, which required the fixing of new steel plates to strengthen the original wrought iron. The cast-iron column capitals were repaired or replaced. The external coverings of the roof were replaced with profiled steel sheeting and a polycarbonate glazing; for reasons of weight and economy the ridge-and-furrow glazing was not restored. The roofs were redecorated in a scheme of white, grey and red, taking a cue from historic schemes without attempting exactly to replicate any one of them.[6]

So by 1990 the station was looking better than it had since before the Second World War, but there was a sense that it was in danger of being swamped by the rising tide of traffic of all kinds: the station's context and the way it was used had changed beyond recognition. Paddington's problems originate in part with the mindset of its first builders and the dilemma they faced in the 1830s. First, the dilemma. Brunel designed a magnificent station, but the site was a compromise and almost certainly not his ideal choice. Initially he had thought about a site at Paddington or maybe the south bank near Vauxhall, but the GWR's first parliamentary bill had the line ending in Pimlico near Vauxhall Bridge. The second bill and the Act of Parliament said that they would share a terminus with the London & Birmingham Railway at Euston. When that idea broke down (*see* chapter 2), they were driven to the present site: sunk in a cutting, hemmed in by the canal and surrounded by residential streets. The tracks into Paddington have a narrow 'throat' which represents a serious constraint, compounded by the limited platform lengths within the station.

As to the mindset: the GWR's main line, like most of the other early trunk routes into London, was initiated and planned by people from the provinces, from Bristol more than from London. They wanted a line to get their trade into London: the same was true of their competitors, the Liverpool and Birmingham men who were behind the London & Birmingham Railway. The last thing on anyone's mind was London's own transport needs. This kind of 'long distance' mentality persisted in the GWR, as in its rivals the London & North Western

Railway (LNWR) and the Midland Railway, right up to the formation of the 'Big Four' in 1923 and beyond. They never really developed suburban lines and services out of Paddington: they preferred to use the Metropolitan Railway, which they partly owned, to provide limited suburban services on their behalf. The railway companies to the south and east of London, by contrast, saw suburban services as being at the heart of what they did: the result was the radically different development of railways there – and of the communities they served.

Thus, the GWR probably thought of their station as a way into or out of London, rather than as part of the city's own transport network. Their old route-map is rather telling: the whole vast network fans out to the west and narrows down towards London, with Paddington at the far end of a spur of railway lines which plug it into the city. The company almost certainly never thought of themselves as providing 'commuter services', which would perhaps have offended their corporate pride and sense of regional identity. Nevertheless, by the late 20th century the relentless growth of London's economy and its ever-widening economic reach were catching up with the station, changing the nature of its business almost by stealth, and gradually drawing it into the city's own complex and overloaded infrastructure.

In the 1980s the Thames corridor was growing steadily more crowded and economically important. More and more people were commuting into the station on a regular basis: whether the trains carrying them could technically be described as 'suburban', 'regional' or 'long-distance' services was arguably ceasing to be the point. All these passengers needed to get somewhere else on arriving at Paddington, tipping increasing numbers onto the underground system. As anyone who uses Paddington knows, its underground stations are oddly and rather awkwardly placed, the Hammersmith & City, Bakerloo and Circle Line platforms all being at some distance from each other. So there are complex cross-currents between the various underground platforms, as well as the people getting on and off main-line services.

Greater London's transport problems were engulfing the station in other ways. It is quite hard to get to by road: the string of termini along the Euston and Marylebone roads are much better connected. The opening of the Westway flyover in 1972 poured hundreds of extra cars onto the narrow streets around Paddington,

as well as bringing a new visual blight to the whole area. In the 1980s British Rail and Westminster City Council began to grapple with the problems seriously. British Rail commissioned a master plan from the architects Michael Hopkins & Partners, while Westminster City Council commissioned a study of the area and its planning issues from the engineering practice Alan Baxter & Associates (1990). This was roughly the situation when British Rail was privatised and its network and stations transferred to Railtrack plc in 1996.

By this time two more major issues had arisen, one of them a distant prospect, the other an imminent reality: Crossrail and the Heathrow Express. Crossrail – the idea of a railway line spanning London from east to west – could be regarded as the great missing link in London's transport system, which should have been provided back in the 1830s if anyone had been thinking of railways as part of the metropolis and its infrastructure and not just as a means of getting to it. A Central London rail study of 1989 made the first serious proposal, and its main western stop was going to be at Paddington. For the time being this was shelved, given its vast cost. In 2009 work actually started on this huge project. Its impact on Paddington is considered later: Crossrail, first mooted in the 1980s, is unlikely to be complete and open before 2018.

The Heathrow Express, by contrast, was translated from idea into reality rather quickly, though with hindsight it seems surprising that it hadn't happened at least 20 years previously. Back in 1946, when Harmondsworth Aerodrome was designated by the Ministry of Civil Aviation as London's new big civilian airport (while it was still in Ministry of Defence ownership, in order to avoid a public inquiry or any other kind of democratic scrutiny of the idea), there seems to have been an assumption that road access and the Piccadilly Line would suffice. By the 1980s there was a clear need for a proper rail link: previously Waterloo had been favoured, but by the time a detailed study was made in 1985 Waterloo was judged to be full to capacity already, so the choice fell on Paddington almost by default.[7]

Railtrack commissioned a further master plan for the station from Nicholas Grimshaw & Partners. The impending closure of the Post Office's sorting premises at the back of the Lawn presented an opportunity, while the Heathrow Express idea represented a catalyst.

The work involved realigning platforms 6, 7 and 8, building new platform canopies outside the 'country end' and bringing overhead 25kV power lines into the station. The footbridge was remodelled and a new tunnel formed beneath the platforms. A new taxi canopy and an elegant new glazed link to the Metropolitan Line platforms by John McAslan & Partners were among the associated works. Overhead electrification was introduced into the train shed for the first time.[8]

The next phase of work was a remodelling of the Lawn and the general concourse area at the platform heads to designs by Nicholas Grimshaw & Partners, carried out between 1989 and 1999. The entrance to the underground station was moved forward to allow the Lawn to be paved uniformly and a lot of unsightly clutter to be cleared away. Culverhouse's simple elevations of patent Victoria stone were cleaned and re-glazed and airline baggage check-in facilities for the Heathrow Express created behind them (see Fig 7.31). The Lawn was repaved and provided with shops and amenities for travellers (Fig 6.6). The new mezzanine structures are in fair-faced precast concrete, while the fine new roof, its form reminiscent of the Paxton ridge-and-furrow glazing, has glazing fins, mullions and fixing pieces of sand-cast steel. Grimshaw & Partners had recently built the long and sinuous Eurostar terminal at Waterloo and their

experience there was developed at Paddington, the carefully considered details and their close involvement in the manufacturing processes echoing the engineering culture of Fox, Henderson & Co. A row of train indicators which blocked the long views down Brunel's shed was removed: their replacements are at right angles to the long axis, leaving the view unimpeded. Another welcome gain was the reinstatement of the original end-glazing pattern to the three Brunel spans. These works made the station work more coherently and restored some visual order to it, enabling its spatial qualities to be better appreciated.[9]

The aesthetic gains were very welcome, but the prospect of the Heathrow Express opening in 1997 threatened to tip hundreds more taxis a day into the station. A good half of them were prospectively going to arrive down a picturesque but very narrow Edwardian taxi ramp at the far end of 'Span Four', which still serves as a passenger entrance (a second Edwardian ramp, descending from an entrance in the flank of Span Four, had previously been blocked by the Red Star depot). Arriving by this route, as your taxi rounded a sharp corner, plunged into the cavernous darkness of the train shed and descended a ramp with the trains alongside you, was aesthetically a most satisfying experience, but as (prospectively) the main taxi access for the Heathrow Express it seemed

Figure 6.6
The Lawn area remodelled as part of the 1998–9 refurbishment of the station.
[© English Heritage K010975]

fairly implausible. The Metropolitan Police then resolved the matter by judging that in the era of the car bomb allowing taxis right into the station was a security hazard, and closing the ramp anyway.[10] This left the cramped taxi rank on Eastbourne Terrace, Brunel's old departure side approach. So Paddington was in desperate need of a new vehicle access, and this led everyone's attention back to the Bishop's Road Bridge just east of the station, which was an impressive piece of Edwardian engineering, but had a weight restriction on it and was only two lanes wide. The result was the Paddington LTVA (Long Term Vehicle Access) project, to replace the Edwardian bridge with something stronger and wider incorporating a new taxi access to the north side of the station. Replacing a major bridge over one of the busiest railway lines in Britain presented the planners and managers with yet another problem, or series of problems: this remarkable project and its unexpected historical outcome are outlined towards the end of chapter 8 (*see* pp 151–7).

The station's ownership changed again in 2002, after Railtrack plc were forced into administration by the then Labour Government in the aftermath of the 2001 Hatfield rail crash: like the rest of the network it was transferred to Network Rail plc. Shortly before its demise Railtrack had commissioned designs from Nicholas Grimshaw & Partners for the replacement of

Span Four. The brief for this involved an extra platform, lengthening platforms and providing new taxi access, all to be provided in a massive concrete structure on the north side of the station on the site of the Edwardian Span Four. There was to be a major 'air rights' development on top, to help pay for it all. The proposals were controversial on the grounds of their scale and effect on the historic station, and few can felt any regret when they were withdrawn in 2005.

Instead, Span Four has been magnificently refurbished, the work completed in 2011 (Fig 6.7 and *see* Fig 7.4). The 'Paddington Integrated Project', a joint venture between Network Rail, Crossrail, Transport for London and London Underground Limited, has also been completed, to provide a new vehicle entrance on this side of the station (Fig. 6.8). This is a good example of the complexity of infrastructure projects in a historic city. The Crossrail project displaced the old vehicle entrance on Eastbourne Terrace. The only other place for it was on the north side in London Street, alongside Span Four, but to make this happen a new vehicle ramp had to be built across the Hammersmith & City Line platforms, which needed improvement anyway. So the Paddington Integrated Project had to take place, embracing these works, as necessary preparation for Crossrail.[11]

Crossrail, an idea of Brunel-like grandeur to make an east–west route for main-line trains

Figure 6.7
Span Four, restored and cleaned, 2011.
[© English Heritage DP104693]

Figure 6.8
The 'Paddington Integrated
Project': a model showing
the north side of the station
as it is being rebuilt. A
ramp leads down from the
Bishop's Road Bridge to the
new taxi entrance.
[© Weston Williamson
Architects]

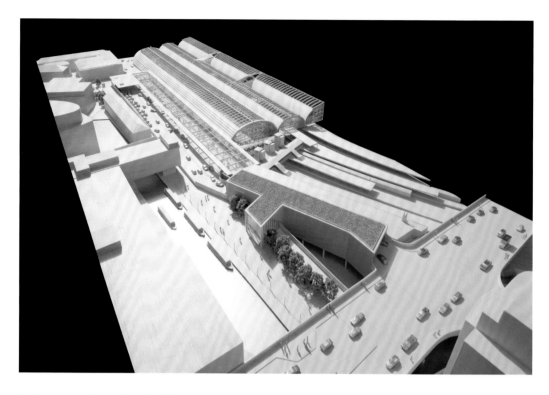

below central London, had been actively canvassed and planned since the mid-1980s. It was envisaged from the outset that this would link Paddington to Liverpool Street, but this was clearly going to involve truly formidable engineering and architectural questions. Early Crossrail proposals envisaged taking the new bored tunnels right under Brunel's train shed and the hotel, but serious doubts arose about the feasibility of all this. It turned out that the effect of such a scheme on Brunel's great sheds could not be calculated and that the tunnel would interfere with the piled foundations of the 1930s extension behind the hotel. The scheme was revised to take the Crossrail line beneath Eastbourne Terrace. A Crossrail bill was at last brought before Parliament in 2005, and the Act received royal assent on 22 July 2007. Despite concerns about the almost £16 billion cost, construction began on the line at Canary Wharf in May 2009. The platforms beneath Eastbourne Terrace will be linked to the passenger station by a new concourse on the site of the present cab ramp, which will receive a new glass canopy at higher level. The new Bishop's Road Bridge and the renovation of Span Four will tie Paddington better into its street context, allowing it to breathe more easily. Crossrail, when it opens in 2017–18, will transform Paddington yet again, tying it much more firmly into London's own transport network. Further thousands of passengers will be added to the great numbers who use the station every day (Figs 6.9, 6.10 and 6.11).

Paddington has had a narrow escape: the previous scheme for the Span Four site would have crushed the station visually under the weight of its high-rise towers, replaced the Edwardian span with a concrete basement ceiling and taken the morning light from the train shed, marring it more than anything has done yet. The disappearance of this threat makes the refurbishment of Span Four, completing the restoration of the magnificent train sheds, all the more sweet: the station seems set fair for the 21st century. For vast numbers of passengers who use the station every year, an arrival at Paddington will continue to provide moments of visual pleasure to lift and lighten the stressful, workaday business of travel, as they step out under that broad and welcoming roof.

Paddington is one of the world's great railway stations and one of London's finest Victorian monuments, but it is not a museum. It remains one of the busiest stations in Britain, as essential to the nation's economy as it was in the 1850s. There used to be a widespread view of Victorian architecture which regarded buildings like the Crystal Palace as the harbingers of a bright modernist future and argued that reactionary Victorian society had ducked this challenge, preferring instead to hide the 'truth'

Figure 6.9
Crossrail: an artist's rendering of the Eastbourne Terrance range as it will be, with the new Crossrail entrance canopy, 2011.
[© Weston Williamson Architects]

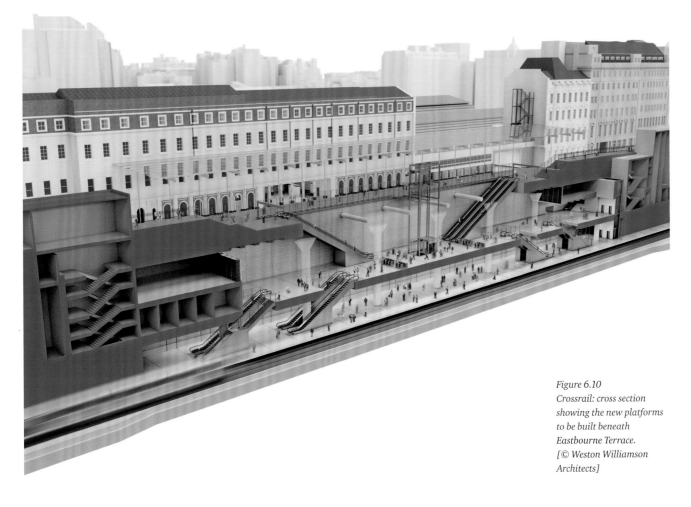

Figure 6.10
Crossrail: cross section showing the new platforms to be built beneath Eastbourne Terrace.
[© Weston Williamson Architects]

Figure 6.11
The Crossrail project will convert the old taxi ramps on Eastbourne Terrace into the station's main entrance for pedestrians, as seen in this computer-generated image. The large new structure to the left will house emergency ventilation for the Crossrail tunnels.
[© Weston Williamson Architects]

Brunel's achievements were on a heroic scale, but as Angus Buchanan has noted, as engineers have tamed the natural world, so their achievements have gone from seeming heroic to appearing merely workaday.[12] The 'heroic age of engineering' has passed into history: no engineer today would have the opportunity to do what Brunel did. He had plenty of vested interests to contend with, including parliamentary committees, avaricious landowners and a board of directors with a ferocious attachment to economy, but he did not have to operate within our overcrowded modern landscape with its inflated property values. Nor was he trapped inside the labyrinth of local plans, planning controls, amenity groups, multi-modal studies, public inquiries, health and safety regulations, environmental impact assessments, conservation legislation and political control over budgets that contemporary architects and engineers inevitably and rightly have to contend with. Some observers have liked to remark on the 'modern' qualities of the great Victorian engineers, but in truth the men who built the railways needed different qualities to the primarily political and bureaucratic skills that are the key to success in most contemporary engineering projects. Today, managing the politics, the finance and the dense forests of planning problems are the really difficult part: the technical problems of building the railway or bridge tend (with distinguished exceptions like Crossrail and the new Bishop's Road Bridge) to be straightforward by comparison.

Paddington was the product of a very different culture to our own, harsher and less democratic but more elevated and serious-minded, in which railways were too new and represented too much effort to be taken for granted and were instinctively seen as objects of corporate and national pride. Brunel and Stephenson were popular heroes; the GWR was nicknamed God's Wonderful Railway: would anyone invent such a nickname today?

Our culture has celebrities, not heroes, and very few of the celebrities engage in anything as useful as engineering. It is hard to see how our society could allow any one architect or engineer the opportunity to display the synoptic boldness of the early railway builders. But one may at least hope that we can once again value our railways as they deserve, for their history and their immense and continuing contribution to society, and to rekindle some of the pride in them that once found such superb expression in Paddington Station.

of their industrial culture behind a veneer of historicist decoration. Applying this kind of thinking, one might well assume that an engineer like Brunel would have had little time for ornament on his buildings, this being added only to humour the taste of the board or the first-class passengers. In fact, as we have seen, Brunel arranged the involvement of Matthew Digby Wyatt and Owen Jones without telling the board and the whole of Paddington's rich decoration was smuggled in by them under the directors' noses and against their explicit opposition. For Brunel and Wyatt their station was a work of art as well as of engineering: its cathedral-like spatial qualities were appropriate to its status, as was the peculiarly original ornament which they devised for it. It was a sad loss that Owen Jones' colour scheme was never carried out to complete the design. At Paddington one senses the Victorians' justifiable pride in their technology, coupled with a feeling that colour and ornament still had important roles to play, and that without them, industrialisation and mass production might make the world sterile and uniform.

7

The Architecture of the Station

Paddington Station is a masterpiece, but it is not the masterpiece Brunel intended to build (Figs 7.1 and 7.2). His original vision of 1850, dimly glimpsed in his sketchbooks with its central entrance from Conduit Street (now Praed Street), would have expressed its layout and structure much more clearly to the outside world, in the way that Lewis Cubitt's contemporary King's Cross does. This idea was scrapped

by the board's wish to give the Praed Street frontage to the hotel, and Brunel conceived his executed design as an interior: at that time it was probably the largest single-roofed space in London after the Crystal Palace, 'in a cutting, and admitting of no exterior, all interior and all roofed in … .' The broad central trackbed separated the departure side and the arrival side, and the shape of the train shed reflected these

Figure 7.1
Paddington Station in 2013 showing planned alterations for Crossrail.
[© English Heritage]

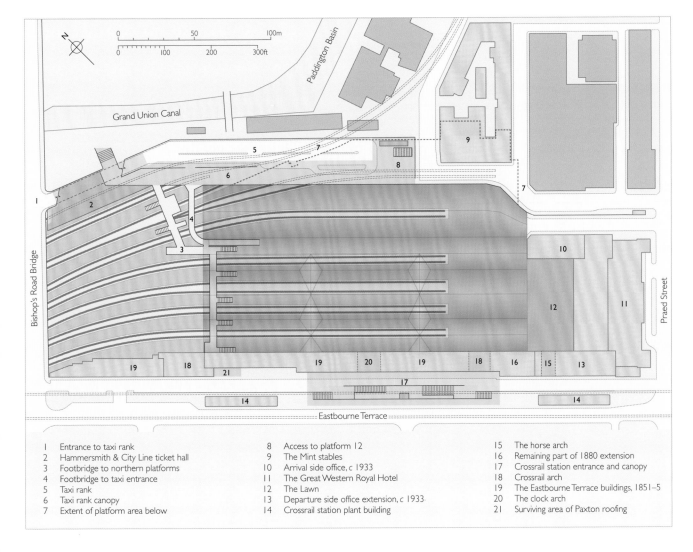

1	Entrance to taxi rank	8	Access to platform 12	15	The horse arch
2	Hammersmith & City Line ticket hall	9	The Mint stables	16	Remaining part of 1880 extension
3	Footbridge to northern platforms	10	Arrival side office, c 1933	17	Crossrail station entrance and canopy
4	Footbridge to taxi entrance	11	The Great Western Royal Hotel	18	Crossrail arch
5	Taxi rank	12	The Lawn	19	The Eastbourne Terrace buildings, 1851–5
6	Taxi rank canopy	13	Departure side office extension, c 1933	20	The clock arch
7	Extent of platform area below	14	Crossrail station plant building	21	Surviving area of Paxton roofing

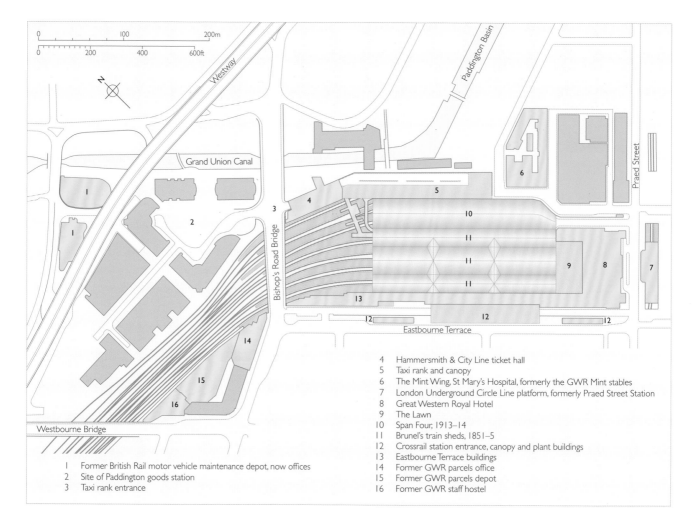

4   Hammersmith & City Line ticket hall
5   Taxi rank and canopy
6   The Mint Wing, St Mary's Hospital, formerly the GWR Mint stables
7   London Underground Circle Line platform, formerly Praed Street Station
8   Great Western Royal Hotel
9   The Lawn
10   Span Four, 1913–14
11   Brunel's train sheds, 1851–5
12   Crossrail station entrance, canopy and plant buildings
13   Eastbourne Terrace buildings
14   Former GWR parcels office
15   Former GWR parcels depot
16   Former GWR staff hostel

1   Former British Rail motor vehicle maintenance depot, now offices
2   Site of Paddington goods station
3   Taxi rank entrance

Figure 7.2
Paddington and its environs
in 2013.
[© English Heritage]

functional divisions. The original layout and its operation are discussed in chapter 5.

The train shed

Brunel's great roof remains largely as built (Fig 7.3), other than the loss of its original Paxton roofing. When it was new this was the largest train shed in existence, the main roof measuring about 700ft (213m) by 240ft (73m). Of the three spans, the central one is 102ft 6in. (31.2m) wide, that to the south is 68ft (20.7m) wide and that to the north 70ft (21.3m) wide. To the south the shed is enclosed by the immensely long departure platform or platform 1 façade. To the north, the shed originally opened into a lower aisle covered with ridge-and-furrow Paxton roofing: the site is now occupied by the fourth span of 1909–14 (Fig 7.4 and *see* Fig 6.7).

The shed's immense length is broken up by the two 'transepts' or transverse axes. There has

been a persistent story, originating in Isambard Brunel Junior's biography of his father, that they were designed to allow for the installation of traversers across the width of the station, but it seems clear that no traversers were ever built in these positions. There is some evidence that the transepts were designed to allow for the installation of 'descending platforms', sections of platform measuring 40ft (12.2m) by 25ft (7.6m) designed to be lowered on hydraulic arms, but it is not clear whether these were ever installed or what their purpose would have been (*see* chapter 5). The transepts, in any case, fulfil a structural function in providing longitudinal stability to the roof (ie they resist any tendency for the long rows of arches to 'rack', or topple over sideways): we have no documentary evidence on the subject, but it seems likely that this was Brunel and Fox's intention. The concept may well have been carried over from the design of the Crystal Palace where an arched transept performed a similar structural

Figure 7.3 (above)
Brunel's triple-span train
shed looking west from the
main passenger concourse.
[© English Heritage
K010987]

Figure 7.4 (left)
The entrance to Span Four,
as refurbished in 2009–11.
[© English Heritage
DP104682]

purpose. Visually, the transepts are a vital element in Paddington's interior, giving it a richness of spatial composition which made the design and manufacture of the roof more complicated and added to its cost, so there must have been a practical reason for including them. The reason seems to be long forgotten, but visually they are a vital element in Paddington's architectural quality, giving it a spatial quality which, as Henry-Russell Hitchcock remarked, is 'almost unique in Victorian railway stations'.[1]

The shed is divided into bays by rows of columns. Counting from the outer end of the station, the rhythm is: 'country end' – 7 bays – transept – 6 bays – transept – 7 bays – 'town end'. Each bay is spanned by 3 arches, making 63 in all (a total of 189 over the 3 original spans). Only every third arch sits directly over a column, the others being carried by X-braced girders of wrought iron, which span from

Figure 7.5
A view along the arches,
showing Matthew Digby
Wyatt's applied cast-iron
ornament.
[© English Heritage
K030802]

column to column and are about 30ft (9m) long.

The most immediate precedent for the design of Paddington's roof was Newcastle Central Station, designed by the architect John Dobson and built in 1847–50, the ironwork by the founders Hawks Crawshay (*see* Fig 3.4).[2] Newcastle originally had three parallel sheds, each about 60ft (18m) wide, with curved roofs sitting on arches, and here too, every third arch sits over a column, the intermediate arches being carried on the girders. The major differences are that at Newcastle, the difficult site was turned to advantage by laying it out on a curved plan; the arches have a segmental profile; and they have tie-rods.

In his sketches for Paddington, Brunel envisaged his roof as having segmental arches too. In the event it was built to a revised design, with five-centred arches which give the design more visual 'lift' than segmental arches would have. This crucial change to the design probably came from Sir Charles Fox and Fox, Henderson & Co, though it is not like anything they had previously executed (Fig 7.5). Fox didn't give the Paddington roof tie-rods, which would have interrupted the long views down the sheds. Instead, the arches are seated on 'skewbacks' – upright sections with a sloping face which sit back-to-back above the columns. The outward thrust of each arch is transferred into the skewback, and balanced by the arching forces of the next span over. The central span, being wider, exerts a greater thrust than the side spans so the forces aren't quite equalised, which has tended to push the rows of columns slightly out of the vertical. Span Four was given tie-rods, making it into a form of sickle or crescent truss during its construction, for similar reasons (*see* p 121 and note 55).

The arches are made of wrought-iron plate, 16in. (406mm) to 18in. (457mm) deep. Most of Fox, Henderson's previous iron roofs were carried on triangular trusses made up of relatively light wrought-iron members, sometimes of wrought iron and timber in combination. The Crystal Palace of 1850–1 had been roofed with flat wrought-iron trusses on a module of 24ft (7.3m) for the aisles and 72ft (21.9m) for the main nave: it was possible for Fox Henderson to build it in five months because of the limited number of components and the modular simplicity of the design.

Paddington's wrought-iron arches would have required more elaborate preparation.

They could have been built of wood: Paxton's 'Great Stove' glasshouse at Chatsworth was carried on laminated timber arches, as was the transept of the Crystal Palace. Brunel's largest station roofs, at Bath and Bristol Temple Meads, were both roofed in timber, and Lewis Cubitt was building King's Cross Station with laminated timber arches around the same time. Brunel had built laminated timber arches over 80ft (24.4m) long for his 'Skew Bridge' to carry the railway over the Avon at Bath. So it would probably have been possible to roof Paddington in timber too, but Brunel does not seem to have entertained the idea.

The principal models for the kind of wide-span iron roofs that Brunel was sketching in late 1850 were Decimus Burton and Richard Turner's Palm House at Kew (1845–7), John Dobson and Hawks Crawshay's roof at Newcastle Central (1847–50) and Turner and Locke's sickle trusses at Liverpool Lime Street Station (1849–50). At Kew, Turner had started with 12ft (3.7m) rolled I-section ships' beams of wrought iron, forged them together, and worked out how to bend them to the right curvature to make 50ft (15m) arches. Turner developed this technique for his even bolder 153ft (46.6m) roof at Liverpool Lime Street, where he fitted ships' beams, just 9in. (229mm) deep, together to form great segmental arches, trussed with cast-iron struts and wrought-iron rods: this seems to have been first crescent or sickle truss. The iron founder William Fairbairn, who was advising him, rightly insisted on tests: the first test models failed.[3] At Newcastle Central, the closest model for Paddington, Hawks Crawshay were making wrought-iron arches with a 60ft (18.3m) span: they found it impossible to bend I-section beams at this scale, and instead they built up their arches from riveted lengths of wrought-iron plate, with L-sections attached to make the top and bottom flanges.[4]

At Paddington, as at Newcastle, the arches are made up of sections of rolled wrought-iron plate, as is shown on the contract drawings.[5] The webs (the up-stroke of the 'I') are made of several sections, ¼in. (6.4mm) thick, bolted together. They are all curved to make the arch shape: we do not know how Fox, Henderson made them to the correct curvature. The flanges are made of lengths of wrought-iron plate, L-shaped in section, bolted to each side of the web at top and bottom. A subtle point is that the flange pieces diminish in thickness, from ¾in. (19mm) to ½in. (12.7 mm) to ⅜in.

(9.5mm) as the arches rise. Again, we do not know how they were made, but they certainly required great skill in ironworking.

The complete arches would have been too large to travel, whether by train or barge. So the components were probably sent from Smethwick (they probably came by barge down the Grand Junction Canal, as the GWR's line to the Midlands was not yet open) and riveted together on site. We do not know whether the arches were assembled on the ground and lifted into place, or whether they were built up piece by piece from a scaffold.

All the arches have round and star-shaped perforations in their webs – 'stars and planets'. These do not appear on Fox, Henderson's contract drawings, and were probably contributed by Brunel or Matthew Digby Wyatt. Visually, they are an ingenious idea making an important contribution to the design in the long raking views.[6] They also have a practical purpose, to carry scaffolding for painting and maintenance work. The arches also carry decorative panels of cast-iron tracery, bolted to either side of the webs low down. These were designed by Matthew Digby Wyatt and do not appear on Brunel's sketches or on Fox, Henderson's contract drawings (Fig 7.6).[7] In addition to the long, leaf-like tracery sections at the arch springings, which still exist, there used to be numerous smaller tracery sections higher up, probably masking the joints in the arches. Most of these seem to have been removed at some point in the 20th century, and now they can only be seen on the external façades. In 1988 the remaining tracery sections were found to be causing a serious problem because moisture trapped against the ribs was mixing with sulphur dioxide and other pollutants and eroding the ironwork. The solution adopted in 1988–91 was to fix steel reinforcing plates to the webs and to bolt the decorative cast-iron elements back over these.

The central nave, from the level of the tracks to the underside of the ribs at their apex, is 54ft 7in. (16.6m) high; the arches, from the springing point of the ribs to their apex, are 33ft (10m) high.[8] The roof was originally supported on 69 identical cylindrical cast-iron columns, given flared bases and bolted to blocks of York stone set in heavy concrete foundations. Their appearance is known in detail from the original drawings, and also from Frith's famous painting (see Fig 4.2). Matthew Digby Wyatt probably designed their curious and original

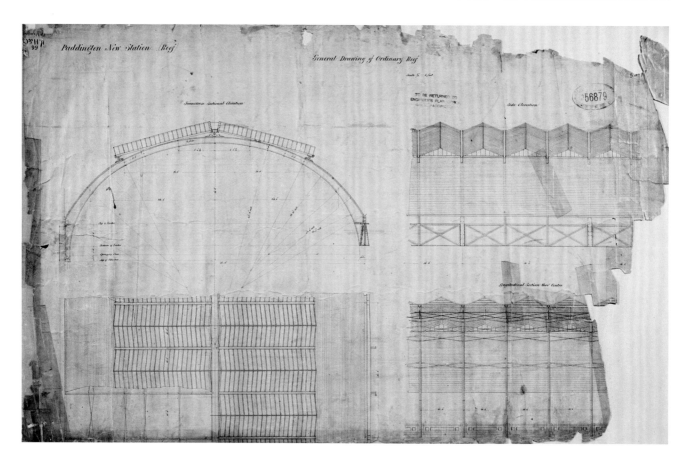

Figure 7.6
Fox, Henderson & Co's
design for the main roof,
1851.
[Network Rail/EH
AA031088]

ornament: the capitals branch out with tall lugs or 'handles'. In 1914 the row of columns on modern platforms 9 and 10 was replaced in riveted steel to support the weight of the new fourth span and the rest of Brunel's columns were replaced in like manner in 1922–3.[9] The new columns are octagonal and have a very different effect. A couple of original cast-iron columns survive at the 'country end' of the shed, one plain cylindrical one and a decorative four-lobed column with a leaf capital at the end of platform 1. The latter, a unique survival, was one of those designed to carry the roof over the Lawn area and has been moved.

The columns carry the longitudinal girders, formed mostly of simple L- and X-shaped sections. The intermediate arches come down onto 'skewback' sections, which sit back to back and form the end uprights of the girders. The original designs show an oak packing between the skewbacks and the arch bearings, presumably to allow for movement in the structure, and perhaps also to allow a degree of tolerance in construction. The girders transmit the load of the intermediate arches sideways: as the architect Nicholas Grimshaw has observed they must be the most heavily stressed elements in

the building.[10] The weight of the intermediate arches is expressed visually by simple timber bosses suspended below the girders: they have no structural purpose.

Paddington's roof is notable in not having purlins – that is, beams which run longitudinally to carry rafters and the roof covering, and to provide lateral stability. Brunel and Fox were able to do without them because their arches are so closely spaced. Instead, the roof has X-shaped braces made of flat wrought-iron straps, between the arches. The lower part of each span was originally covered in corrugated galvanised iron sheeting, a popular new material for industrial buildings (and is covered with an equivalent modern material today): Brunel seems to have had some involvement in this decision.[11] The profiled cladding runs longitudinally, so it provides the lateral stability in conjunction with the X-shaped braces. Paddington is thus an early example of what engineers call a 'stressed skin' design: the roof covering does not just sit there, it makes a positive contribution to the stability of the building.

The upper part of the shed roofs was original covered with 'Paxton roofing' (Figs 7.7 and 7.8). This was a system of ridge-and-furrow glazing

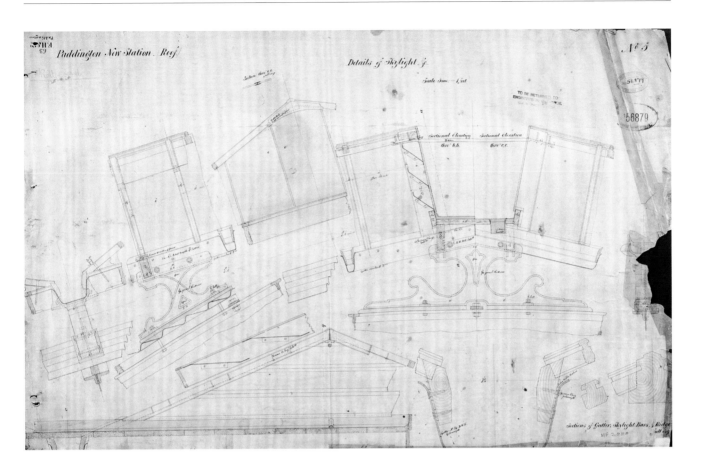

Figure 7.7 (above)
Fox, Henderson & Co, a large-scale detail of the glazed Paxton roofing for Paddington, 1851.
[Network Rail/EH AA031093]

Figure 7.8 (left)
The Paxton roofing being cleaned, 22 April 1932. The Bishop's Road Bridge appears in the background as rebuilt in 1906. The Bishop's Road Station is visible to the right.
[Fox Photos/Getty Images]

made of plate glass and wood, originally developed by Joseph Paxton at Chatsworth in 1849–50 (*see* chapter 3). The glassmaker Richard Chance of Smethwick had introduced the French technique for cylinder glass into England in 1832, refining it to produce sheets up to 3ft (0.9m) long. Paxton cajoled him into developing the technique further, to make 4ft (1.2m) sheets. With this, he developed his system of glazing, buying a Boulton & Watt engine to cut the timber glazing bars for the 'Great Stove' of 1836–40. In 1849–50 Paxton perfected his design for a new greenhouse to house the giant Victoria lily: ridge-and-furrow glazing using 4ft-long sheets of Chance's 26oz (737g) glass set in timber frames, with the gutters in each valley also serving as the structural support, and the structural columns serving as drainpipes. Evidently, it was a design which could serve in many different situations, and Chance persuaded him to apply for a patent. The Crystal Palace generated a need for Paxton roofing on an industrial scale, probably sooner than they expected.[12] Shortly afterwards, it was used at Paddington: we do not know if the GWR or Fox, Henderson & Company paid Paxton a fee, but as the patent holder, he would surely have been entitled to one. The glass almost certainly came from Chance Brothers again, whose works were almost adjacent to Fox, Henderson & Company's 'London Works' in Smethwick.[13]

The glazing has been renewed on many occasions: the board minutes for *c* 1905–15 contain numerous references to the overhaul of the main roof. It was removed entirely in 1968–9 and replaced with simpler flat-profiled glazing.

The major renovation of 1988–95 involved the replacement of the roof coverings: the upper slopes were given flat-profiled glazing in a polycarbonate material. It was felt that renewing Paxton roofing would put too much weight on the structure, given its age. It was an understandable decision, but the loss of the Paxton roofing is regrettable, for historic reasons and also visual reasons: it gave the roof a richer and more complex geometry.

Each span is closed in and glazed at either end. The gable ends were all given an absolutely simple glazing, of rectangular panes. Just inside these, there are screens of decorative tracery made of wrought-iron straps, forming an 'anthemion' pattern on a huge scale, presumably designed by Matthew Digby Wyatt. Most of this had been destroyed in the 20th century but was restored on the basis of the original drawings in the course of works of 1988–95.[14]

As we have seen, Brunel wanted his station to have ornament, and commissioned Matthew Digby Wyatt to supply it. Henry-Russell Hitchcock wrote of their work at Paddington that 'there seems to be so coherent an integration between the general handling of the interior space and the detailing of the structural elements that a true collaboration between Brunel and Wyatt on the over-all design must, I think, be assumed'.[15]

The ornament is eclectic and in marked contrast to the simple 'engineers' classical' of most early railway buildings. The columns have a vaguely Moorish quality, but Professor Hitchcock also related them to the 'overlappings and penetrations one finds in the detail

Figure 7.9
One of Brunel's sketch designs for Paddington, showing his first idea for the oriel windows overlooking the transepts, early 1851. [University of Bristol Brunel Collection (Large Sketchbook 3, 28). By courtesy of the Brunel Institute, a collaboration of the SS Great Britain Trust and the University of Bristol]

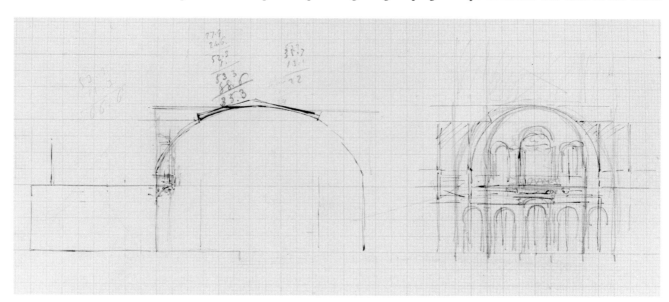

Figure 7.10
'Entrance to Offices,
Paddington Station'.
Matthew Digby Wyatt
worked up Brunel's sketch
into this design, as built
from The Builder, *17 June*
1854
[No593, Vol XII, p323]

of E B Lamb' (a notably original mid-Victorian architect, who delighted in abrupt and startlingly detailed ornament). The tracery on the end glazing, might again be considered Moorish but it might equally be thought of as based on Classical 'anthemion' patterns. The panels of cast-iron tracery applied to the main roof trusses are Gothic in character, with a certain quality of *Art Nouveau avant la lettre* about them.

The façade to platform 1 is a symmetrical composition, though it is so long that this can hardly be appreciated. In the centre is the five-bay entrance loggia, marked out by its square-headed openings and by the huge clock projecting above. To the east and west there are impressive set-piece compositions overlooking the two transepts of the train shed, which were illustrated by *The Builder* on 17 June 1854 (Figs 7.9 and 7.10). These have fine oriel windows in cast iron at the upper level, lighting the very long corridor which served

the GWR offices. Below these were doorways – that to the east leading to the first-class waiting room, that to the west leading to the royal waiting room. The façade's decoration is more or less unclassifiable. Hitchcock wrote of it, 'if it is not iron but cement, Wyatt must have consciously assimilated the modelling to the character of ironwork'.[16]

In fact the fat roll mouldings and the faceted bosses, here and on the façade to Eastbourne Terrace, are all of cast iron: on some of them the scratched foundry numbers in Roman numerals can be made out. Wyatt may have simply used sections of iron guttering, applying them to the wall as decoration. The filigree spandrel panels about the arches and the strapwork panels at girder height are of cast iron too: they look at first as if they might serve some practical ventilation purpose, but apparently do not.

This strange and original detailing reaches its height in the two fantastic oriel windows looking over the transepts and the lace-like

Figure 7.11
The GWR's war memorial,
unveiled on Armistice Day
1922.
[© English Heritage
K010984]

Owen Jones' bright colour scheme, commissioned by Brunel and Wyatt, was vetoed by the GWR board. Jones was a pioneer of colour theory, applying his ideas to the Crystal Palace: yellows advance; blues recede; reds are suitable for the middle distance; white for vertical elements.[17] Sadly, no evidence of his design for Paddington has ever been found: if it followed the same principles as his scheme for the Crystal Palace, we might imagine it as having white columns with white and yellow capitals, the arch webs painted blue with white flanges, the tracery against them yellow and white, the longitudinal girders red and white. In the event, *The Builder,* describing the completed train shed, said 'colour is sparingly introduced – red and grey'.[18] This has broadly been confirmed by paint scrapes made by English Heritage, which found several 19th-century paint layers but with shades of white, grey, ox-blood red and a kind of salmon pink predominating. It is possible that what *The Builder* saw in 1854 was incomplete, for an engineer's report of 15 February 1855 said that the painting had 'been designedly postponed until next summer'.[19]

The royal waiting room entrance was redesigned in 1919 by Thomas Tait of Burnet, Tait & Lorne to accommodate the GWR's very fine war memorial. Its centrepiece is the magnificent bronze figure of an infantryman by Charles Sargeant Jagger (Fig 7.11).[20] The doors to either side are 20th century and presumably contemporary with it.

By 1900, the broad trackbed which originally filled the middle of the shed had been filled in with alternating tracks and platforms. These have been remodelled on several occasions to suit the company's (and later British Rail's) changing requirements. The original platforms had timber surfaces and very probably rested on timber structures. These have been replaced in masonry and concrete, at unknown dates and probably in several campaigns.

Eastbourne Terrace – exterior

When Brunel wrote to his friend Matthew Digby Wyatt in January 1851, he described the station as 'in a cutting, and admitting of no exterior, all interior'. There is some slight exaggeration here, and evidently he was not counting P C Hardwick's hotel as part of the station. One sees his point, however: Paddington had no demonstrative façade and entrance to compare with the London & North Western's at Euston

ornament around the edge of the roof here. Henry-Russell Hitchcock discussed the characterfulness and slight ungainliness of Wyatt's work, not realising that his designs were based on Brunel's sketches. Their work at Paddington may be seen as a step in the invention of an architectural vocabulary for ironwork, and in some respects presages the work of some of the most energetic High Victorian 'rogue' architects such as E B Lamb, or R L Roumieu and A D Gough.

or the Great Northern Railway's at King's Cross. The one major façade was to Eastbourne Terrace (Fig 7.12). When built it was only two storeys high, immensely long and low, and set about 15ft (4.6m) below street level. It seems clear that Brunel was basically responsible for its design, not Wyatt. In January 1851 he wrote to Saunders, the company secretary, ' "At least you shan't say I don't try to adopt others' suggestions. I have been for some time past trying to see if I could not adopt your plan of putting the offices over the range of buildings of the booking office, and the two new features" – an inclined plane parallel with Eastbourne Terrace and a glazed roof covering thereto … .'[21]

All of these ideas feature prominently in the buildings as they developed and as they are today. The 'inclined plane' that Brunel

referred to has not changed appreciably since its construction in 1851–2. It was originally paved in small granite blocks.[22] Towards Eastbourne Terrace there is a retaining wall that carries heavy cast-iron railings. Several original designs for the railings survive, including a number of full-size drawings (Fig 7.13).[23] These were altered in the late 19th and early 20th centuries, and mostly replaced after wartime bomb damage: today, the original railings only survive to the west of the 'cab roof'.[24]

Brunel made some attempt to articulate this very long and plain façade, without achieving overall symmetry. The central block was 32 bays long, with a 20-bay centre and six bays at either end projecting forward very slightly. The ground floor was given a decorative arcaded treatment, with round-headed sash windows

Figure 7.12
Paddington Station and Eastbourne Terrace c 1922, showing the houses opposite, many of them occupied by the GWR.
[Great Western Trust Collection]

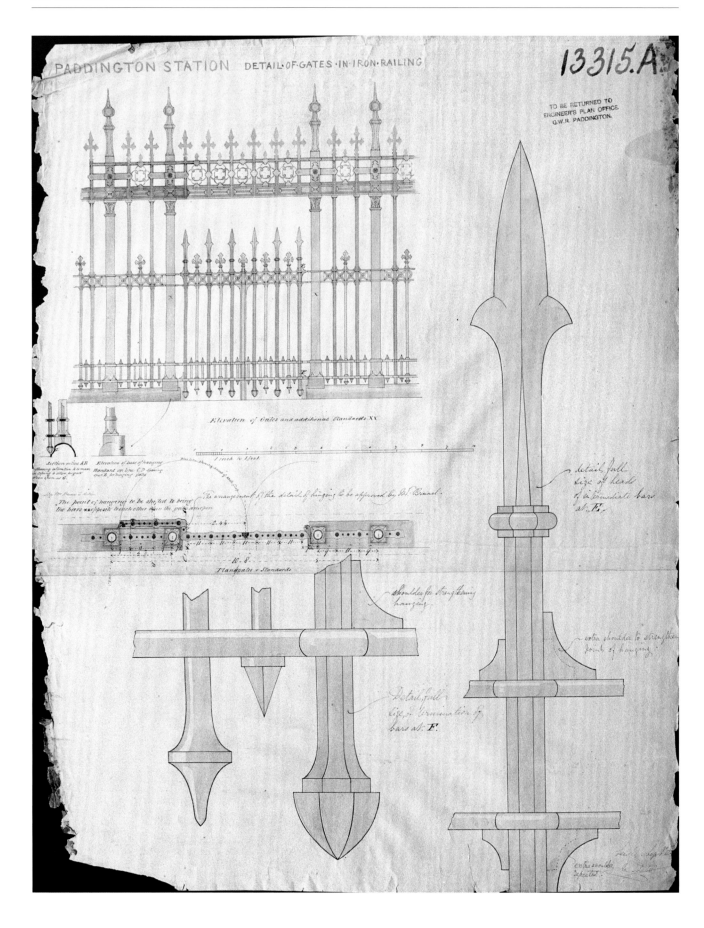

surrounded by rounded mouldings similar to those inside the station.[25] The upper floors were (and are) plain and stuccoed, with tall sash windows. Network Rail's archives have a large undated elevation for the middle of the Eastbourne Terrace façade, showing a more elaborate decorative scheme with arches, window aedicules on the first floor and a bracketed cornice (Fig 7.14).[26] This was probably designed by Wyatt but was never executed, doubtless because of the need for haste and economy.

Near the centre of this main 32-bay section is the loggia or vestibule which forms the main entrance through to platform 1 – the original departure platform. As built, this had five bays to the platform but seven bays to the outside: the difference was accounted for by quadrant-shaped bays, which were filled in the early 20th century and have been partly reinstated. The loggia's ceiling originally had decoration in Elizabethan style strapwork with pendants of wood and plaster, but this has been covered over with modern tiling: the original decoration does not seem to survive.[27] Intrusive light fittings and a crude modern statue of Brunel

do little to adorn this space: Paddington and Brunel deserve better.

At ground-floor level, the outer seven bays at either end of this section are stressed with ornament. Each of these sections is centred on a large doorway, that to the west originally leading to the royal waiting room, that to the east, originally to the first-class waiting room. The royal waiting room survives but the first-class waiting room was destroyed in the Blitz.

When the buildings were new, the range jumped back by about 20ft (6m) to the west of the 32-bay centre and stretched westwards for another 20 very plain bays, with simple arcading framing round-headed windows on the ground floor. This took the building to just beyond the west ('country') end of the train shed. There was a break in the buildings, with a vehicle entrance to the departure platform and milk platform covered by a long run of Paxton roofing. Beyond lay another detached block, 15 bays long, of stock brick with stucco window surrounds. In 1876 the gap was filled with another 14 bays of plain stock-brick building.[28] There are six bays of original Paxton roofing

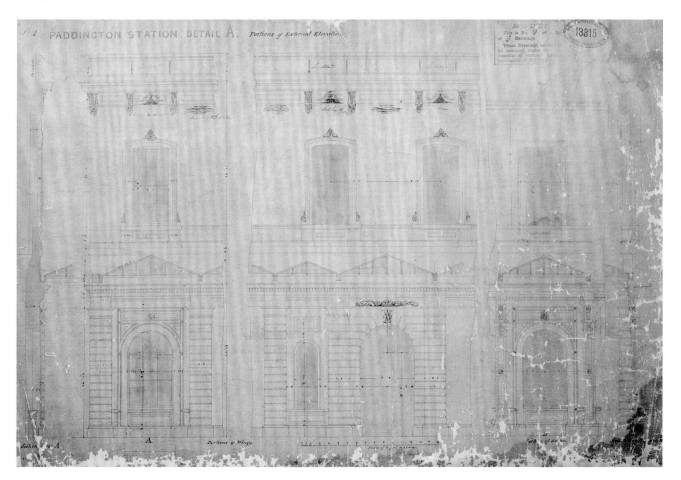

over the roadway here: neatly repaired in 1990, they are the only example of this style of roofing to remain at the station, which used to be largely covered with it (Fig 7.15).[29]

To the east of the 32-bay centre, there was a short spur of four bays and then a long gap as far as the south end of the hotel: here, there was a vehicle entrance to the platforms (the 'horse arch'), and then a high, blank screen wall concealing the train shed. The gap was filled in stages, as the GWR's need for office space grew. In 1880–1 the four-bay spur was doubled to eight bays and a higher and more ornate nine-bay block added to the east. This made more of an architectural statement, being faced in Greenmoor stone with blind arcading at ground-floor level, window aedicules above,

a deep bracketed cornice and a high mansard roof.[30] At ground or platform level the new block housed a second booking hall and a way through to the departure platform. A separate entrance led to a big new shareholders' meeting room on the first floor, and the upper floors housed the engineer's department. The high mansard roof space housed the GWR's plan store until it was moved to Swindon in the 1960s. The shareholders' meeting room block was partly damaged by the parachute mine in 1941, and its western two bays were among the areas taken down afterwards.

In the 1930s the offices were extended further to the east as part of the great bout of alterations carried out by the GWR's own architect P A Culverhouse. He built a steel-framed extension, 115ft (35m) long and six storeys high, faced in granite lower down and with patent Victoria stone above. It is the most explicitly Classical of Culverhouse's additions but is quietly detailed, with pedimented window surrounds on the first floor and a bracketed cornice.[31] This filled about half of the space between the 1880 building and the hotel and Culverhouse built a one-storey building spanning this gap. A few years later, this one-storey extension was raised to six storeys in a matching style, producing a continuous row of six-storey buildings to Eastbourne Terrace.[32]

In 1879–81, the whole range of buildings was given a second floor of offices, along with a top floor, though much of this seems initially to have been left as empty loft space for future expansion.[33] This huge operation was handled in the same contract as the south-eastwards extension by the same contractors, Kirk & Randall of Woolwich. It involved major strengthening works to the original 1850s building. Numerous drawings show the iron columns and beams that were inserted for this purpose.[34]

There has always been a glazed roof over the central section of the approach road. Brunel was more explicit in a further letter of January 1851, referring to the long retaining wall between the cab road and Eastbourne Terrace, which '... would carry a glass roof over the centre of the road approach in front of the Booking Offices – it would be anything but ugly towards Eastbourne Terrace, and I should think could not be objected to, it is only 12ft [3.7m] high as on the footpath and all open ... by this means we could light the lower storey without skylights ... .' (Fig 7.16).[35]

Figure 7.15
A design for the canopy over the milk platform approach, Eastbourne Terrace, showing how it incorporated gutters with the columns serving as drainpipes. [Network Rail/EH AA042114]

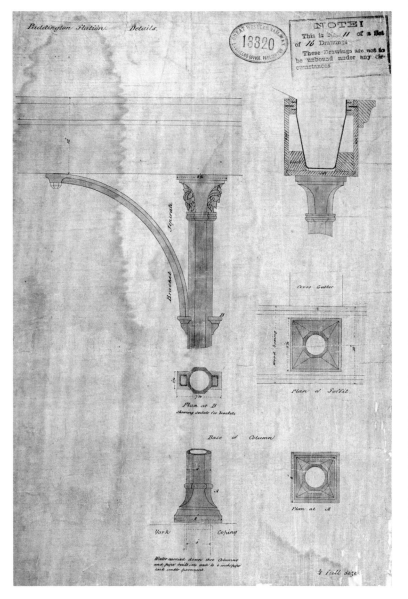

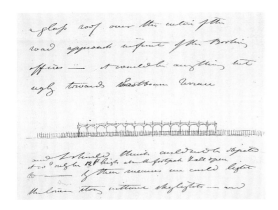

Drawings for the original roof are extant.[36] Brunel used Paxton roofing again, carried on slender wrought-iron trusses. Originally, it was carried on cast-iron columns towards Eastbourne Terrace, with elaborate railings between them.[37] The original roof had Paxton glazing concealed at the 'town end' by a gable with a clock and the GWR arms. In 1880–1 the cab roof was extended in two stages, so that it came part-way along the tall shareholders' meeting room block.[38] It was partly destroyed by the 1941 parachute mine, though some of the original ironwork may survive in the present structure (Fig 7.17). Beneath the repaired cab roof, the entrance to the royal waiting room remains as built: a doorway flanked with Doric pilasters, with a crown supported by the lion and unicorn above, forms the centre of a seven-bay composition. The original railings survive to the north-west of the canopy: the rest seem to be modern replacements, from the repairs after the wartime bomb damage.[39]

The building of the new Crossrail platforms for Paddington will see major changes for this side of the station. The whole of the street (Eastbourne Terrace) will be dug out to build a gigantic concrete box housing the platforms and lines: from here, Crossrail's main tunnels will run eastwards. The 'departure side' road with its ramps will become a pedestrian entrance, the taxis re-routed to the other side of the station. The present roof canopy (which more or less replicates Brunel's design and incorporates some original ironwork) will be replaced with a larger glazed canopy roofed at higher level, turning the area into a new pedestrian concourse which links the new Crossrail platforms at low level to the main-line station (*see* Figs 6.9 and 6.10).

Figure 7.16
Brunel sketches the design for the cab roof on Eastbourne Terrace in a letter to Charles Saunders, 23 January 1851.
[© PRO Rail]

Figure 7.17
The cab roof and the entrance façade towards Eastbourne Terrace, photographed in 2001. Under the Crossrail scheme, the area will be radically changed and the old cab roof replaced with a new, higher glass canopy.
[© English Heritage K010988]

Eastbourne Terrace – interiors

When the buildings were new, the ground floor was occupied (starting at the east end) by a vehicle entrance to the departure platform, a few small offices, the first-class ladies' waiting room, the first-class waiting room, the main booking office, the central loggia, the second-class waiting room, the royal waiting room, smaller offices and then another vehicle entrance. This was fairly close to Brunel's initial sketch plans made in 1851 (Fig 7.18). Drawings of the general office interiors and waiting rooms are dated August 1851 and on 4 December Brunel laid plans of the board and committee rooms before the directors.[40] Fitting out the interiors took rather longer, for some of the surviving full-size detail drawings are dated 1855.[41] The booking office was the most important space, nine bays long, with windows to the departure platform and the approach road, though given the station's low situation they may not have admitted that much light. Perhaps

for this reason Brunel gave it a roof light rising through first-floor level, with a glazed gallery looking down into the hall.[42] This handsome feature was lost in 1880–1 when the extra storeys were added to the building (Fig 7.19). The whole of this side of the range was gutted after bomb damage in 1941.

The western half of the range escaped bomb damage and a little more of the interior survives. Going westwards from the main entrance vestibule, there was a room for train guards and a large cantilevered staircase to the upper floors which survives largely in its original condition. Beyond was the second-class waiting room, later converted into a refreshment room, handsomely modernised by Culverhouse in 1923 with marble columns with brass capitals and mahogany panelling (Fig 7.20). This has been subdivided but it is possible that some of the original decor survives.[43]

From the outset, Brunel intended that the new station should have a separate royal waiting room. A separate entrance under the cab

Figure 7.18
'Booking office buildings according to Plan No. 3', 5 January 1851. Brunel sketched the first plans for the Eastbourne Terrace office buildings himself. [University of Bristol Brunel Collection (Large Sketchbook 3, 9). By courtesy of the Brunel Institute, a collaboration of the SS Great Britain Trust and the University of Bristol]

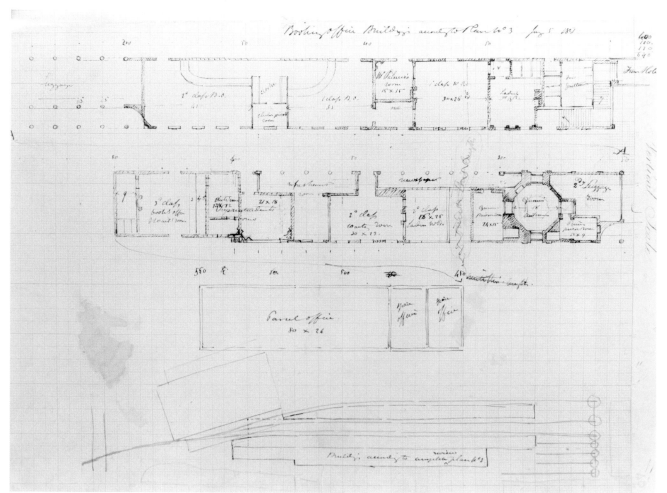

Figure 7.19
The season ticket office and
second-class booking office,
c 1905.
[© National Railway
Museum/Science and
Society Picture Library]

Figure 7.20
The refreshment room as
remodelled in 1923.
[© National Railway
Museum/Science and
Society Picture Library]

roof leads to a lobby, into a main domed octagonal room, with an oblong room opening off it. Originally, another door led straight through onto platform 1 on an axis with the entrance door, but this was blocked by the GWR's war memorial in 1919 and another entrance made using an archway to one side instead. It would seem that, in 1851 or 1852, M D Wyatt asked the sculptor Alfred Stevens to produce designs for the waiting room, though the board seem to have been completely unaware of this. Stevens produced some sketches for a splendid Italianate design, but in the climate of economic difficulties and with the directors determined not to spend money on frills, this never stood any real chance of being executed (*see* Fig 3.20).[44] Instead, Messrs Holland, who were decorating the hotel, tendered to decorate and furnish the room for £1,900, and were instructed to proceed on 10 November 1853. They gave the room stencilled decoration in a vaguely Florentine manner, part of which has been uncovered. Early in the 20th century, the room was redecorated in a simpler French manner and this scheme largely survives (Fig 7.21). After the Second World War, the waiting room was closed and declined into office use. In 1985 it was refurbished for use by business-class passengers. West of the royal waiting room in the late 19th century there was a bookstall, a

Figure 7.21 (above)
The royal waiting room in the late 19th century.
[© PRO Rail]

Figure 7.22 (right)
The long first-floor corridor, the heart of the GWR's head office, c 1905, displaying some of the company's collection of historic views.
[© National Railway Museum/Science and Society Picture Library]

servery and then a dining room, modernised by Culverhouse in 1923, now also converted to office space.

On the first and second floors were the GWR's principal offices. Long corridors ran along the train shed side. The first-floor corridor windows are for the most part above the line of the shed roof, but where the two transepts rise up, the oriel windows provide dramatic views over the train shed. As the company grew, more and more office space was needed. The building was extended at either end and in 1878–80 a whole upper storey was added. After 1880, two corridors 850ft (259m) long ran the whole length of the building at first- and second-floor levels (Fig 7.22).

From the 1850s, the first floor of the centre of the building housed the GWR's senior officers. Over the central vestibule there was the chairman and the secretary; to the west a waiting room and the deputy chairman; to the east, clerks, the committee room, boardroom and luncheon room. There were offices purposely designed for the solicitor, the general manager, the accountant, the 'superintendent of the line' and for T A Bertram, Brunel's resident engineer. Numerous drawings survive for fitting out offices with bookshelves and plan chests as well as standard interior joinery.[45] The work was plain but of good quality: the boardroom and directors' luncheon room were given a slightly more decorative treatment with dado-height panelling and fine double doors.

After the 1880 expansion, the second floor of the central range was occupied by the committee's secretariat, the general manager's office, the library and the accountants. The top floor was also occupied by the engineers, who had their main 'parliamentary drawing office' here. The GWR's main plan store was in the mansard roof above (Fig 7.23): this vast collection of plans and drawings was later moved to Swindon, and is now in Network Rail's national records centre in York. The western end of the buildings, as we have seen, was originally a detached block, the gap being filled in the 1878–80 bout of works. The ground floor of the far wing housed the deeds office and the GWR's printing press, where timetables, posters, leaflets and tickets were produced by the thousand. Above were the finance and audit departments. The 'infill' building had the milk platform entrance at ground level (Fig 7.24) and above that the registration, telegraph and estates departments. When this immense range

of buildings was full, the GWR started renting or buying houses on the other side of the street: by the early 20th century it occupied 21 of them.[46]

An inventory of works of art of 1926 shows how far these rooms had become a GWR company museum. The boardroom was hung with

Figure 7.23 (left)
The GWR's plan store. This great collection, including most of the designs for the station reproduced in this book, was housed next to the chief engineer's department in the Eastbourne Terrace buildings.
[STEAM – Museum of the Great Western Railway, Swindon]

Figure 7.24 (below)
The milk platform at the far end of platform 1, with the outer end of the Eastbourne Terrace buildings c 1914. Behind the barred windows was the GWR's printing department, where all their tickets, timetables, regulations and posters were produced.
[STEAM – Museum of the Great Western Railway, Swindon]

portraits of the great founding figures of the company, including 'the Picture' of Charles Russell and a bust of Brunel, and a great map of the GWR network on the long wall (*see* Fig 5.15). The adjacent directors' room had a copy of Frith's *The Railway Station*, engravings of towns in the GWR's empire and the seals of the 160 or so companies which had been incorporated into it (Fig 7.25).[47] The shareholders' meeting room housed the 'Elkington Shield', the prize awarded for the company's Rifle Corps annual shooting competition, a memorial to the company's servants who had been killed in the Boer War. It was also lined with photographs of company's senior managers, commemorating successive secretaries, general managers, engineers and the like. The immensely long corridors were lined with paintings, watercolours, engineering drawings and models. Sadly, most of the collections seem to have been dispersed after nationalisation in 1947.[48]

At the outbreak of the Second World War the GWR's head office was evacuated to Beenham Grange in Berkshire. In April 1941 a large section of the range including the boardroom was destroyed by a parachute mine. The damaged area was demolished, and eventually the 'bomb gap' was filled with low, modern building that made no attempt to fit its surroundings. Under British Rail, the range became known as Macmillan House, but since privatisation the railway has moved out, and most of the upper-floor space is let as commercial offices. Repeated bouts of renovation have removed almost every trace of the Great Western Railway from the interiors which were their headquarters. A few spaces retain a degree of historic character, notably the royal waiting room, the staircase nearby, the shareholders' meeting room on the first floor of the 1880 extension and elements of Culverhouse's 1923 refreshment room opening off platform 1. Recently, a suspended ceiling was taken down in a big first-floor room at the far ('country') end of the range, revealing original cornice and ceiling, with large decorative iron vents for the original gas lighting.[49]

Figure 7.25
The directors' smoking room, c *1905. Easy chairs, mahogany panelling and 18th-century topographical prints.*
[© National Railway Museum/Science and Society Picture Library]

The London Street Deck and Span Four

When Brunel built Paddington, the northern side of his station was the 'arrival side'. North of his main train shed, a further broad aisle was covered by a flat-profiled roof with Paxton glazing. An entrance ramp ran down from Conduit (now Praed) Street in more or less the same position as it does now. North of the station ran London Street: this divided, with one branch leading down a ramp and beneath the Bishop's Road Bridge into the goods station, the other leading onto the bridge itself. On the far (north) side of London Street was the GWR's coal depot, and north of this, the Grand Grand Junction Canal and its Paddington Basin. All of this is clear from early plans of the station, though few early photographs of the area have been found.

In 1863 the Bishop's Road Station was built for the Metropolitan Railway, involving a change to the line of London Street, and in 1878–82 there were relatively minor alterations: the addition of more waiting rooms and a new milk platform abutting the north-western corner of the station.

The arrival side was reconstructed in the first really major alteration to Brunel's station, c 1908–14. The station had to remain operational, so the complex works were undertaken in several phases, starting with the reconstruction of several of the bridges over the main lines (see pp 150–2). In 1908–9 London Street and the old high-level coal yard were closed. Most of this area was excavated to track level, and a massive new concrete retaining wall built to the north (towards the canal and the Hammersmith & City Line tunnel), 10ft (3m) thick, 15ft (4.5m) high and 550ft (168m) long. To create storage space, over a large part of the area the excavation was taken down deeper and a large basement created, spanned with rolled-steel joists and brick jack arches. At its south end this housed the new concourse for the Bakerloo Line. It also provided storerooms for lost property, the newspaper business, and linen and clothing for the station staff. It also housed the stores for the hotels, refreshment rooms, restaurant cars and steamboat catering department, distributing everything from groceries to linen and crockery around the company's empire. This sequence of basement rooms and corridors, lined with glazed tiles, still exists. Above the new basement two

trackbeds housing four lines (numbers 9–12) were built. The northern one (12) was planned for deliveries of milk, fresh fish, meat and vegetables and the like. To enable milk and other fresh goods to be taken directly off trains and onto wagons, a raised platform was made (number 12) with a sunken area of roadway on the other side, so the milk churns could be rolled straight across.[50]

At the same time, the south end of the Mint stables was demolished, its site dug out to platform level, and a further new basement built in Hennebique ferroconcrete, with the board marking and chamfered beams characteristic of that technique.[51] This housed a new parcels office at track level, with hydraulic lifts to take parcels and sacks up to new vehicle bays, in the reconstructed south wing of the stables. All of this still stands and is visible on London Street, though the vehicle bays have been bricked up.

In 1909–12, a massive steel-framed deck, over 400ft (122m) long and up to 80ft (24m) wide, rose over the new platforms 11 and 12, to carry the re-routed London Street. The riveted steel columns and girders support brick jack arches to carry the re-routed London Street. The engineering is on a very large scale: the columns, some of them weighing 18 tons (18.3 tonnes), sit on steel mesh 'grillage' foundations 16ft (4.9m) square to spread the load. On the north side of the new London Street Deck a steel-framed ramp was made, the 'milk ramp', to take vehicles down to the new milk platform (number 12). Light wells were incorporated in the southern side of the deck to admit light and air to the platforms below: they are identified in early photographs by their palisades of surrounding railings.[52]

The new Span Four roof went up in 1913–15. It was designed by the GWR's chief engineer William Armstrong, the steelwork made by the Horseley Bridge & Engineering Company, and constructed by Messrs Holliday & Greenwood (Fig 7.26).[53] The site tapered at the east end, so the new span could not be quite uniform in shape (Fig 7.27). The roof has 62 ribs, which line up with those of the Brunel roof but are made of steel instead of wrought iron, the webs tapering from 2ft (610mm) depth at the springing point to 1ft 10in. (559mm) at the crown of the arches. The contract drawings show that Armstrong designed them as five-centred arches, and gave them the same decorative perforations of 'stars and planets' as the Brunel spans.[54]

Figure 7.26
Span Four, built c 1913–15.
This image was taken by the
well-known architectural
photographer Eric de Maré,
c 1960.
[Reproduced by permission
of English Heritage
AA98/05430]

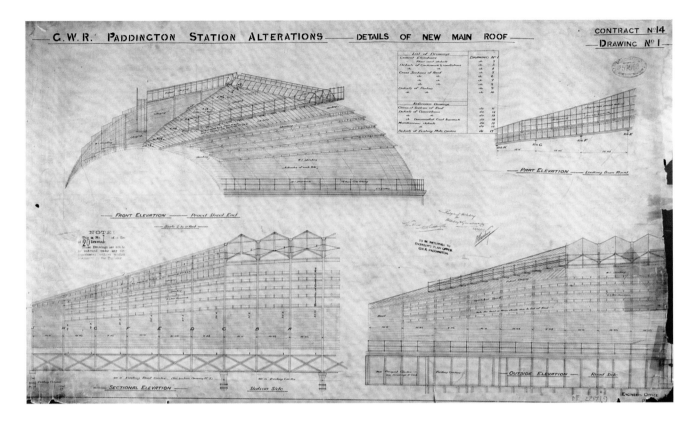

Figure 7.27
The design for the tapering
south-east end of Span Four,
1914.
[Network Rail/EH
AA031094]

The contract drawings show that the new roof was designed without tie-rods, but while it was still under construction the management commissioned an engineers' report on the station roof as a whole. They recommended the addition of tie-rods, which can be seen in photographs of the roof under construction.[55] The roof was covered in the same style as the Brunel roofs, the lower parts covered with corrugated sheeting, while the upper part was glazed with 'Rendle's patent glazing', not far in appearance from the glazing of the 1850s. The north flank of the new span to London Street is faced with panels of sheet steel, the riveting of which makes a rhythmic pattern, capped by a dramatically projecting cast-iron cornice and gutter. Finally, two features at the 'town end' of platforms 11 and 12 deserve notice: two splendid sets of Edwardian hydraulic buffers made by Ransome & Rapier of Ipswich remain *in situ*. The guard railing around them is formed of reused pieces of bull-nose rail and broad-gauge rail: both buffers and railings are precious survivals.

Two more vehicle ramps were built to bring hansom cabs (and later, taxis) into the arrival side. One curved round the far end of the train shed, delivering the taxis to a ramp which descended into the northern span of the Brunel roof between platforms 8 and 9 (this replaced an 1860s ramp). The other entered through the northern flank of Span Four, and descended between platforms 10 and 11 (Fig 7.28). These still stand, though they are no longer in use by vehicles.

The railway tracks and the London Street Deck had to be extended to the north-west in 1930–4, to allow for the rebuilding of the Bishop's Road Station, taking the suburban and Hammersmith & City services. The additional tracks meant that there was a wider gap to span at the north-west corner of Span Four: this required the construction of the largest steel girder that had then been built: a massive bowstring girder weighing 126 tons (128 tonnes), which rises above the street level.[56]

The Edwardian milk ramp seems to have become redundant with the opening of the big new goods depot in the 1920s (*see* pp 134–6). The GWR was developing its own fleet of motor lorries, and created a depot for them on the old coal yard. In 1932–3 A one-storey transport service building was built above the milk ramp, with its petrol tank taking up part of the void created by the ramp.[57] This was demolished at some point in the post-war period: it is not clear when. In 1985–7 British Rail built a new metal-clad shed to house its 'Red Star' parcels depot,

Figure 7.28
A traffic jam of taxis
at the taxi entrance on
the north side of Span
Four, January 1936. The
huge hog-backed girder
rising above the roadway
was installed c 1932–3,
spanning an area where the
tracks of the Metropolitan
and Hammersmith & City
were being re-routed to
run diagonally, requiring
the removal of columns: at
126ft (38.4m) long, it was
the largest single girder ever
made at the time.
[© National Railway
Museum/Science and
Society Picture Library]

which abutted the northern side of Span Four: this was demolished again in 2007.[58]

For many years Span Four was in a neglected condition, concealed by internal scaffolding and a protective 'crash deck'. Around 2005–6, proposals were developed for the demolition of this whole side of the station, and its replacement with a massive concrete structure, forming a low ceiling over remodelled platforms below and supporting a very large air rights development, with tall office blocks overshadowing the station. The proposals aroused considerable controversy, and the conservation pressure group SAVE Britain's Heritage launched a public campaign for Span Four's preservation.[59] In 2008 the scheme was abandoned, partly for logistical reasons to do with the impending Crossrail project on the other side of the station. Instead, the Paddington Integrated Project is almost complete at the time of writing (2013), with Mott MacDonald

as engineers and Weston Williamson as architects. The new Crossrail station has displaced the old vehicle access in Eastbourne Terrace. An access ramp from the Bishop's Road Bridge (*see* pp 151–2), has been built to bring vehicles to the north side instead, with a new entrance, escalators down to platform level, and new entrance for the Hammersmith & City Line platforms. Span Four itself has been cleaned back to its steel frame, refurbished and re-clad (*see* Figs 6.7 and 7.4). This very welcome scheme, an intelligent refurbishment and adaptation of the Edwardian building, has opened the station up to the north. Combined with the new Crossrail concourse it will fit the station for another century, tying it better into the street pattern round about. The renovation of Span Four has given a new lease of life to one of the last of the great Victorian and Edwardian train sheds, enhancing the magnificent spatial qualities of the station as a whole.[60]

The Lawn

This part of Paddington is said to derive its name from having been the stationmaster's garden in the period of the temporary station at the Bishop's Road Bridge between 1837 and 1850. This area was lost when Brunel built the present terminus between his great train shed and the back of the Great Western Royal Hotel. Brunel excavated it out and laid tracks up to the back of the hotel. A simple ridge-and-furrow Paxton roof was built above and iron galleries led from the back of the hotel to the departure and arrival platforms to north and south.

By 1900 the tracks had been curtailed. The area was levelled up and paved as a vehicle entrance. It remained in this state until 1930–4, when it was entirely reconstructed by P A Culverhouse, as described earlier. He lined the area with the simple Classical façades of patent Victoria stone which survive today. A new steel and glass roof was built, with three main spans carried on eight columns. Around the perimeter of the area, was a lower ceiling, with flat decorative rooflights originally glazed in white and amber diffusing glass (Fig 7.29).[61] The interiors behind Culverhouse's simple façades have been stripped out and renewed on several occasions.

The whole of the Lawn area was rebuilt between 1995 and 1998, to designs by Nicholas Grimshaw & Partners. In the process a new steel and glass roof was installed at a much higher level, Culverhouse's simple patent Victoria stone façades were cleaned and several later accretions removed. This has re-emphasised the Lawn's quality as a concourse for the station (Figs 7.30 and 7.31).

Figure 7.29
The Lawn area, decorated for Christmas, 1933.
[© NMPFT/Daily Herald Archive/Science and Society Picture Library]

Figure 7.30 (right)
The Lawn area as redesigned
by Grimshaw & Partners,
photographed in 2001.
[© English Heritage
K010995]

Figure 7.31 (opposite)
The Lawn, showing
P A Culverhouse's simple
Classical architecture.
Briefly adapted to house
the 'Heathrow Express'
check-in facilities, it is now
surrounded by shops.
[© English Heritage
K010976]

Arrival side offices

This handsome Art Deco block of five storeys over a platform-level ground floor, was built in 1932–3 as part of the 1930s redevelopment of the station by P A Culverhouse. It stands on the site of the 1851–4 parcels office. The building has a steel frame and concrete floors, and is distinguished by the tiers of very shallow oriel windows, a motif which Culverhouse used elsewhere on the GWR estate.[62] The words 'G.W.R. PADDINGTON' high up on the attic storey (see Fig 4.18), were originally lit by shell-shaped uplighters. At platform level the block housed the station buffet, at mezzanine level there was the GWR's advertising and publicity department, on the first and second floors the superintendent of the London Division, and on the upper floors the chief goods manager's department.[63]

The Great Western Royal Hotel

The Great Western Royal Hotel at Paddington, designed by Philip Charles Hardwick and built in 1851–4, has considerable claims to historic importance in its own right. It was the first hotel in London to be a major and deliberate architectural statement and at its opening was hailed as the 'largest and most sumptuous hotel in England'. It marked a crucial development in scale and opulence for hotel architecture.[64] It was also the first major example in Britain of what would become known as the 'Second Empire' style (Fig 7.32; see also Fig 3.10).

Hardwick's design is a notably early exercise in the kind of French-influenced building which became popular in the 1860s and 1870s. Its determining features, the tall pilaster bands cum quoins, the bracketed cornices, heavy window aedicules, tall mansard roofs, corner towers with prominent pavilion roofs and chimneys, are part of the vocabulary of the 'Second Empire' style. Henry-Russell Hitchcock pointed out that Hardwick was actually anticipating the Second Empire here.[65] This kind of commercial architecture became so widespread that the building's impact is today rather blunted.

Hardwick's watercolour views of the main elevation and the grand coffee room were exhibited at the Royal Academy in 1851.[66] Where the passenger station was concerned, the GWR board generally tried to resist Brunel and Wyatt's introduction of architectural ornament, but where the hotel was concerned they gave Hardwick a lot more leeway, presumably feeling that it was needed if the hotel was to establish itself as the desirable and fashionable destination they had in mind. Hardwick engaged the sculptor John Thomas, who had made relief sculptures for his Great Hall at Euston, to carve the fine pediment sculpture with its allegorical figures of Commerce attended by Plenty and the Four Continents, in a rather more chaste style than the rest of the façade would suggest (Fig 7.33).[67] The façade was originally a good deal more ornate than it is now, with caryatids and atlantes and stucco garlands decorating the centre, and four copies of the Warwick Vase on the cornice above.

The Illustrated London News published a long description of the new establishment on 18 December 1852.[68] It had 112 suites of bedroom and dressing room and 15 private sitting rooms, most of them part of suites. At one end of the ground floor was a suite of club rooms with a separate entrance. The grandest interior was the coffee room, which set a new standard in hotel and restaurant design. It rose through two storeys, measuring 59ft (18m) by 30ft (9.1m), and 27ft (8.2m) high, and was decorated in a lavish Louis XIV style with marbled columns and caryatids supporting the ceiling (Fig 7.34).[69] Most of the hotel's decoration and furniture were provided by Messrs Holland, but several suites were decorated by one Herr Remon.

The Great Western Royal was the first really grand, monumental hotel in Britain, setting a new standard in size, amenities and comfort. It was also influential in introducing a new kind of commercial opulence into Victorian architecture, its French-inspired 'Second Empire' manner becoming the most popular style for

Figure 7.32
The Great Western Royal Hotel, a late 19th-century view.
[Reproduced by permission of English Heritage BB47/878]

Figure 7.33
The pediment sculpture,
Commerce Attended
by Plenty and the Four
Continents, John Thomas,
1852.
[© English Heritage
B904246]

hotels in Britain: the high mansard and pavilion roofs made for a distinctive skyline, and also housed the servants' accommodation. The Grosvenor Hotel at Victoria Station by James Knowles (1860), the Queen's Hotel at Birmingham New Street Station by Livock (1864) (demolished), the Grand Hotel at Scarborough by Cuthbert Brodrick (1863–7) and the Hotel Cecil in London by Perry & Reed (1885–97) (demolished) are prominent later examples.

In 1931 the hotel was remodelled by Culverhouse. The exterior was greatly simplified, with the atlantes and caryatids and much of the ornament removed from the middle, as well as the fine iron and glass canopy.

The lower storeys were refaced with a granite plinth and the patent Victoria stone which Culverhouse used elsewhere in the station. At the back, towards the station, a new entrance to the Lawn area was made and a major extension to the hotel built. The whole of the ground floor was remodelled and, sadly, Hardwick's great coffee room was subdivided. Culverhouse's work was a typical instance of the 1930s reaction against Victorian taste, but with hindsight it was very regrettable, diluting the hotel's design and weakening its architectural impact. In 2001–2 the hotel was fully refurbished and reopened as the Hilton Paddington Hotel, and remains one of West London's busiest.

Figure 7.34
Philip Charles Hardwick's
design for the hotel's coffee
room, 1851, showing it
largely as built.
[RIBA Library Drawings &
Archives Collections]

Related Buildings and Structures

As the Great Western Railway grew and became increasingly complex, the station needed more and more space for its many supporting operations. Several of these related buildings survive but many, particularly those of the goods and locomotive departments, do not. The passenger station never operated in isolation, and it would be misleading to leave these other buildings out: a summary account of them is given, though in most cases, a lot more could be written.

Metropolitan and Hammersmith & City Line platforms – the former Bishop's Road Station

The original Bishop's Road Station was built in 1863 as the western terminus of the Metropolitan Railway.[1] Bringing the line here was complicated, as the whole area had been developed by this stage, and involved bringing two lines of track from the Metropolitan Railway's tunnel at Edgware Road, beneath South Wharf Road and skirting the corner of Paddington Basin beneath the GWR's coal yard to join the north side of the company's tracks. A brick entrance building went up, approached from the Bishop's Road Bridge (hence the name), with stairs down to the new platforms. These works displaced Brunel's original entrance ramp to the company's coal yard, which had to be rebuilt, carried on a series of brick vaults: these housed stables for the company's goods yard. A passenger walkway carried on iron columns linked the new station to the main shed. The Bishop's Road Station lost its status as the western terminus of the Metropolitan Railway within a couple of years, when what is now the Circle Line was extended southwards in 1865: the line was extended westward too, and the station became a stop on what is now the Hammersmith & City Line.

The Bishop's Road Station was only slightly affected by the Edwardian expansion, but it was completely demolished and rebuilt in the remodelling of 1930–4. Two new tracks were needed, to handle the increase in underground and suburban services. Taking suburban services into the underground network required the steam locomotive (for suburban work) to be replaced with an electric one: the new works also involved creating a 'shunt tunnel' or engine spur, north of the station beneath the coal yard, to house waiting electric units. These works involved creating two new tracks, north of the existing ones, to raise the number of platforms to four. The Victorian Bishop's Road Station was completely demolished, and replaced with simple platforms and canopies, probably designed by the GWR's engineer and architect, Raymond Carpmael and P A Culverhouse, rather than by London Underground's designers. The station lost its separate identity, the new platforms being absorbed into Paddington Station. Today it is platforms 13–16: the Metropolitan and Hammersmith & City Line platforms. The present Paddington Integrated Project will create a new entrance to them from the north (Fig 8.1).

Circle Line platforms – the former Praed Street Station

Praed Street Station (Fig 8.2; *see also* Fig 4.6), just over the road from the Great Western Hotel, was built *c* 1865 as part of the extension of the Metropolitan Railway southwards from Paddington to Kensington, when the road was called Conduit Street. Designed by Sir John Fowler or his office in a similar style to other Metropolitan Railway stations (for example the Metropolitan Line platforms at Farringdon and Baker Street), it was made in a cutting with massive, heavily buttressed retaining walls of yellow brick, using very simple Classical

Figure 8.1 (left)
The Paddington Integrated Project will create a new entrance to the station on its north side, including a new entrance to the Metropolitan and Hammersmith & City Line platforms.
[© Weston Williamson Architects]

Figure 8.2 (below)
An early 20th-century view of Praed Street Station on the Metropolitan Railway, now the Circle Line platforms. Note the central power rails: the line was electrified in 1906.
[STEAM – Museum of the Great Western Railway, Swindon]

Figure 8.3
The station from the west, c 1910. This view can be dated by the works for Span Four in progress on the left-hand side: Note, from the left, the arriving milk churns below and the taxi-entrance arch in the partly built flank of Span Four above (seen in figure 7.28); the signal box and the linemen's huts; the two-storey 1850s staff hostel between them and the main shed; the old-fashioned clerestory carriage; the semaphore signals; and the sparsely peopled platform – the milk churns suggest that this is early in the morning. [STEAM – Museum of the Great Western Railway, Swindon]

ornament. The station retains its original roof, or at any rate the original frame, carried on handsome wrought-iron arches with an elliptical shape, not dissimilar to those in the main Paddington train shed. The façade to Praed Street is of *c* 1925, probably by C W Clark, architect to the Metropolitan Railway from 1921. It is neoclassical, faced in white faience, and very similar to his station façades at Farringdon (1921) and Great Portland Street (1930). The station no longer has a clear separate identity, being simply the Circle Line platforms at Paddington.

The Bakerloo Line platforms

The Bakerloo Line platforms, Paddington's only true underground station, were opened with the Bakerloo extension line from Edgware Road to Queen's Park on 1 December 1913. They were built as part of the fourth span works, the concourse and the escalators being directly beneath the train tracks. The most striking feature of the addition was the absence of lifts and the use instead of 'moving staircases or "escalators"', the largest that have yet been

constructed'. The original escalators have, of course, been replaced with new ones of similar dimensions, but the bold Classical ornament on the archway above, and the doorways to the platforms, remain.[2]

Signal boxes

Information on the station's signal boxes has been very hard to come by: no drawings for them have been found, nor have any references to their construction. The first edition Ordnance Survey 25in. map of *c* 1870, however, does mark them (*see* Fig 4.4). A substantial signal box is marked on the south (departure) side of the tracks, just west of the Westbourne Bridge, opposite the carriage department's workshops. A much smaller signal box is marked just west of the Bishop's Road Bridge towards the north side of the tracks: from its position, this may have regulated Metropolitan Railway traffic into the Bishop's Road Station, rather than main-line traffic. By the third edition of the map (1914–16), a larger signal box had been established on the north side of the tracks, between the Bishop's Road Bridge and the

Bishop's Road Station, probably as part of the 1906–14 redevelopment, and a couple of photographs show this (Fig 8.3). The 1930–5 works seem to have provided two new boxes, both in solid brick, which still survive. The departure side box seems to have been a one-storey wing of the parcels depot building on Bishop's Road of 1932–3. The arrival side box was presumably the free-standing structure which remains on the north side of the tracks, just west of the Bishop's Road Bridge.

Paddington goods station

The history of Paddington's goods station and the host of related industrial buildings which once opened off the tracks is at least as complex as that of the passenger station. Hardly a trace now remains of this great group of buildings and only a short account of them is given here. When Brunel started planning Paddington in 1836 his first 'concourse' design included a goods depot on more or less its eventual site, with a new basin opening off the Grand Junction Canal (see Fig 2.6).[3] The prospects for trans-shipping goods from railway to canal and vice versa were presumably a significant inducement for the GWR to choose Paddington as a site. In the event, as we have seen, these early plans were shelved and for the first 20 years the GWR made do with a timber-framed and timber-clad goods depot on the site of the present passenger station: it was 330ft (100m) long by 120ft (36m) wide, with three parallel roofs.[4]

This was not an ambitious start, but the GWR seems to have been oddly slow to develop its goods and parcels business. In the early years the service was run by carriers, who paid the company 15s (75p) per ton for general goods and 10s (50p) per head for cattle and horses. When the line opened to Bristol, the company put on just two goods trains each way per day: in 1842, the £26,845 in goods receipts represented just 3 per cent of the company's revenue. The company was, however, already beginning to think in terms of the trade in coal and building materials: to develop the former, they encouraged their engineer Daniel Gooch to form a partnership with a Welsh coal company.

By 1850 the GWR's goods service was carrying over 350,000 tons (355,616 tonnes) of merchandise a year, producing £202,978, or 24 per cent of turnover. More accommodation was needed and Brunel reserved a large area northwest of the Bishop's Road Bridge, between the main lines and the canal (Fig 8.4). It was important to link the railway with the canal, but the

Figure 8.4
General plan of the goods station, c 1856.
[© English Heritage]

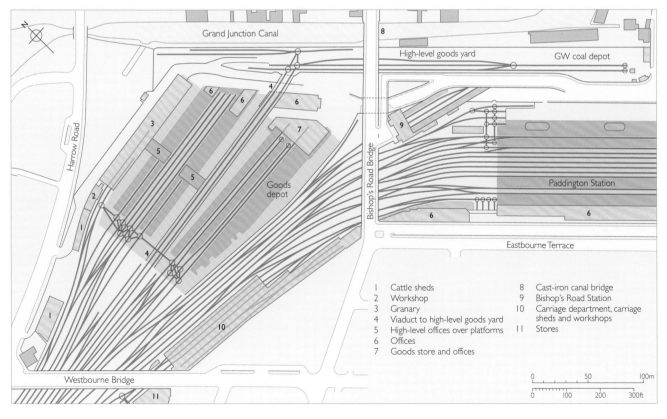

change in level (the canal side is 6m above the line of the tracks) presented a difficulty. Brunel planned a 'high-level yard' alongside the canal: a viaduct led up to this from the tracks. This was primarily used for the company's coal trade, the idea being to get this heavy bulk material up to street level for easy loading onto wagons. After the GWR's lines reached Wales their coal business grew steadily and by 1846 they were transporting 4,350 tons (4,420 tonnes) a month to London. In 1856, shortly after the new high-level yard was fitted out, the GWR signed an agreement with the Ruabon Coal Company (Fig 8.5). The GWR's takeover of the South Wales Railway in 1861 increased the trade further. By 1900 the company was transporting over 1,000,000 tons (1,016,047 tonnes) of coal a year to London. The coal yard was complicated to manage. A difficult junction between the viaduct and the high-level yard tracks was too sharp to be turned by a train, which meant that every wagon had to be uncoupled and turned individually on a turntable, a very inefficient and labour-intensive process: it is surprising that neither Brunel nor his successors installed a curved ramp to eliminate this wasteful feature.

The viaduct to the high-level yard divided the goods depot site in two and Brunel built the larger northern part of it first. A sketch of January 1851 represents his first thoughts for this area: a north–south section through the buildings, it shows how he planned to cope with the change in level.[5] Brunel told the GWR board that the plans for the goods depot

were complete on 12 June 1851 and in July a tender of £23,432 was accepted from Messrs Sherwood.[6]

The higher, northern side of this complex was formed by a very large granary (Fig 8.6), a warehouse four storeys high and 30 bays long. Its lower floor opened directly onto the goods shed at track level. The next floor up opened northwards onto a large yard, entered from the Harrow Road. Detailed drawings in Network Rail's archives show the steel columns and roof trusses for this building. To the south of this a broad area of track and platforms was covered with two 32ft (9.8m) spans of iron and glass roofing, carried on cross-braced trusses with a segmental profile. To the south this was flanked by a narrower roof over a roadway and beyond that another broad segmental-profile roof.[7] Then came the viaduct rising to the high-level yard, which was carried on cast-iron columns: in 1906, for reasons which are not apparent, it too was given a segmental roof and then partly rebuilt in 1909 (Fig 8.7 and see Fig 4.13).[8]

This first stage of the goods station was complete by the end of 1852, allowing the old timber shed site to be demolished, clearing the way for the second phase of the new passenger station. In March 1853 Brunel presented the board with outline plans for the rest of the goods station south of the viaduct, at an estimated cost of £50,000.[9] This was fitted into a confined area between the viaduct and the main tracks. At the east end Brunel planned a warehouse, five storeys high with an irregular four-sided shape. Running westwards from this were four

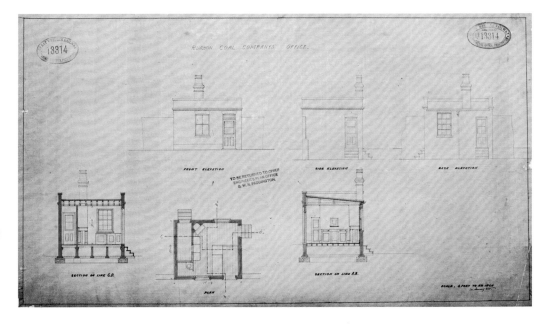

Figure 8.5
An office c 1870, probably sited in the high-level goods yard for the Ruabon Coal Company. It is a measure of the high standards of Victorian draughtsmanship that so much attention could be lavished on this tiny building, hardly larger than a garden shed.
[Network Rail/EH AA031082]

Figure 8.6 (above)
A drawing for the first
phase of the goods station,
showing the granary in
section and elevation.
[Network Rail/EH
AA031052]

Figure 8.7 (left)
The goods depot, c 1910,
looking under the iron
columns carrying the
viaduct to the high-level
goods yard.
[© National Railway
Museum/Science and
Society Picture Library]

trackbeds under a long three-span roof, the central span 76ft (23.2m) wide with segmental tied trusses covered in galvanised iron, flanked by flat spans 41ft (12.5m) wide covered in Paxton roofing. Flanking this to the south was an approach road covered with a 36ft (11m) lean-to roof with raked bowstring trusses, again covered in galvanised iron. Detailed drawings for these buildings are dated June 1854 and August 1855.[10] The new goods buildings were equipped with hydraulic equipment made by Sir William Armstrong & Company at Elswick on the Tyne.[11]

The first goods offices were housed partly in a simple two-storey building at the 'town (east) end' of the goods yard and partly within the northern goods shed, over the tracks.[12] In the great Edwardian rebuilding a huge new range of goods offices was built facing directly onto the Bishop's Road Bridge. Four storeys high and 22 bays long, these were completed in 1906.[13] The goods department had filled the whole available site at Paddington, but the GWR's goods services continued to develop. Having filled the available land at Paddington, they expanded eastwards via the Metropolitan Railway (now the northern part of the Circle Line), opening new depots at Smithfield Market (1869), Poplar Dock (1878), the Royal Docks (1900), and in South Lambeth (1913). To the west, a large new depot was constructed at Westbourne Park (1908). Nevertheless, in the 1920s the Paddington goods depot was still handling 500,000 tons (508,023 tonnes) of general merchandise a year and had 900 road vehicles passing through it every working day.[14] Significant alterations and renovation had been carried out at the time of the 1908–14 works, but in 1925 the company decided to rebuild the goods station completely.[15]

The Victorian buildings were cleared in stages (Fig 8.8; *see also* Fig 4.13), a process recorded by the GWR's photographers: only the long warehouse on the north side remained. The old high-level yard was closed and cleared, the cast-iron viaduct leading to it demolished (Fig 8.9). Huge reinforced concrete cellars were built beneath the eastern part of the site by the Indented Bar & Concrete Engineering Company, then six double-rail bays were laid

Figure 8.8
The goods depot, c 1926. The viaduct to the high-level yard has just been demolished, but its columns are still in place. [Great Western Trust Collection]

out, served by six platforms varying in width from 21ft (6.4m) to 39ft (11.9m). An immense, uniform roof of seven steel-framed spans was built by the Cleveland Bridge Company, with enough clearance to allow for the installation of overhead electric lines in the future.[16] The new shed was designed to house up to 330 wagons (Fig 8.10). The old hydraulic equipment was replaced by new electric plant, a fleet of electric trucks was bought and, in an echo of Brunel's ideas, moveable bridges were installed to span the rail bays, allowing greater ease of movement across the width of the depot.[17]

The goods station remained busy through the 1950s. A proposal by the architect Sergei Kadleigh to build 'High Paddington', an astonishingly ugly Corbusian high-rise scheme, on the goods yard site, came to nothing (*see* chapter 6 and Fig 6.4). However in the 1960s, like the rest of London's goods depots and like the docks, it fell victim to changing patterns of trade. The use of coal fell dramatically with the Clean Air Acts, while deliveries of milk and foodstuffs had overwhelmingly been transferred to road traffic. The goods depot was closed, cleared and sold in fairly short order

Figure 8.9
The former high-level goods yard with its tracks taken up and the goods offices in the distance, 1923.
[© National Railway Museum/Science and Society Picture Library]

Figure 8.10
The new goods depot in use, 1927.
[© National Railway Museum/Science and Society Picture Library]

during the late 1970s. The last major section to go was the 1906 goods offices, demolished in 1986. The demolition contractors, Higgs & Hill, generously arranged for the main entrance doors and doorcase, the mosaic floor from the entrance hall and the giant wooden letters which spelt the words 'Great Western Railway Goods Station' and 'Great Western Railway Goods Office' on the sides of the building, to be given to the Great Western Trust at Didcot.[18] As for the site, the Westway flyover was built over part of it and opened in 1972. The remainder of the site was sold by British Rail, and long remained empty. In the later 1980s, the developer Rosehaugh Stanhope proposed to move the railway station to the goods yard site, converting the terminus for retail purposes, and in 1989 another developer, Regalian, proposed an ambitious development scheme for the area, involving building over the tracks. With hindsight, given the pressures on Paddington from the ever-increasing traffic, it is arguably regrettable that the goods site was not retained in railway ownership for potential use as part of the station. As it was, part of the site was redeveloped *c* 2002 with a large mixed-use development, Sheldon Square, which includes offices, flats, and a hotel. The rest of the site has been covered with a structural deck, but awaits construction of the Crossrail project before it can be developed further.

The locomotive department: Westbourne Park

The GWR's first engine sheds tended to be makeshift timber affairs, but at Paddington Daniel Gooch designed an interesting roundhouse shed, probably the first of this pattern to be built, in 1837–8. The main octagonal structure was 130ft (39m) in diameter and

straight-sided sheds projected from this to the east and west, making an overall composition 360ft (110m) long. Inside the roundhouse, a turntable gave access to eight radial roads. The whole was timber framed and timber clad and only one historic view of it is known (*see* Fig 2.11). As the traffic increased more space was needed and another engine house was built further to the west in 1848.

Brunel needed the site of Gooch's roundhouse for the new goods station and as part of the great redevelopment of 1851–5 he moved the whole locomotive department about a mile to the west, to Westbourne Park (Fig 8.11). An immense brick engine shed was built there, 71ft (21.6m) wide and 663ft (169m) long (for comparison, the passenger station train shed is 700ft (213.4m) long, and Winchester Cathedral, the longest medieval church in Europe, is 556ft (169m) long). It housed four engine tracks spanned with a simple roof of tied wrought-iron trusses, and was built by Messrs Locke & Nesham for what seems the comparatively small sum of £14,130: the contract was signed in July 1852.[19] Parallel to this, just to the north, Brunel planned another range to house offices and workshops. A one-storey range, 11 bays long, housed shops for the smiths, fitters, coppersmiths and a carpenter. At either end were two-storey wings, one with a storeroom below and sleeping quarters for the enginemen above, the other with an office below and a room for Daniel Gooch above (Fig 8.12).[20]

In 1861 a new shed was needed for standard-gauge engines. This was just to the west of the big broad-gauge shed, measuring about 135ft (41.1m) by 45ft (13.7m) and housing three tracks. It was doubled in size to 135ft (41.1m) by 90ft (27.4m) to accommodate another three tracks in 1873.[21] As the broad gauge lost ground, so the bigger shed was gradually converted to the mixed gauge. Nevertheless, the two buildings continued to be known as the broad-gauge and narrow-gauge sheds up to their demolition *c* 1906.

On the opposite, south side, of the main line a broad area was laid out as a mileage yard, with another group of workshops entered from Westbourne Park Road. A coal stage with cranes surmounted by a clock tower straddled the Green Lane Bridge, a site later given over to relief lines for Westbourne Park Station. Finally, a 42ft (12.8m) turntable stood to the west of the broad-gauge shed, believed to be the only one on the GWR at the time that was

Figure 8.11
The locomotive department at Westbourne Park in the late 19th century.
[Great Western Trust Collection]

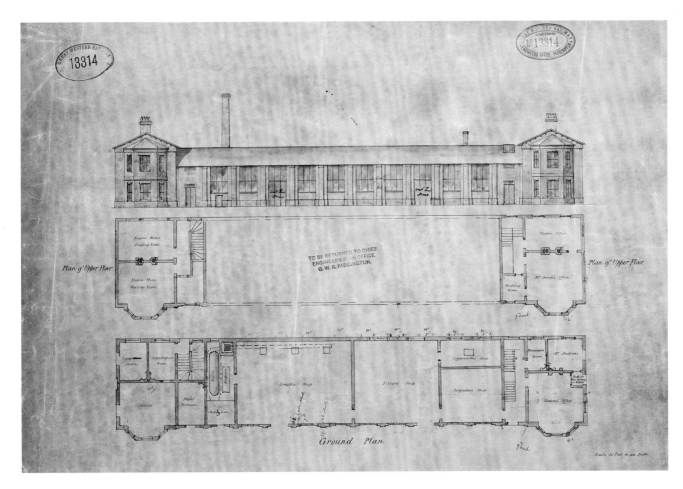

worked by steam: a small boiler house adjoining it provided the power. More or less the whole of this complex seems to have been cleared c 1906, after the GWR opened its new locomotive depot at Old Oak Common, just north of Wormwood Scrubs about two miles (3.2km) to the west.[22]

The locomotive department: Old Oak Common

In 1906 the GWR transferred the main locomotive depot for the London end of the line to Old Oak Common, just north of the main line. Old Oak Common remained in use as a locomotive depot right through the British Rail period, up to its closure in 2009. At the time of writing the site is being redeveloped as the main locomotive depot for Crossrail, where its trains will be serviced and stabled. As this involves the complete demolition of what was one of the GWR's most important depots, a somewhat fuller account than in the first edition seems called for.[23]

In 1902, George Jackson Churchward (1857–1933) became the Great Western Railway's chief mechanical engineer, formally bearing that title from 1916. Churchward set about the modernisation of the company's fleet of locomotives, innovating and investigating foreign ideas. His new engines, larger and more powerful, were again built at the company's Swindon works. They became celebrated for their power, efficiency and reliability, and were a major element in the company's 20th-century revival. To accommodate them, he and his team planned a new generation of standardised depots, mostly built between 1906 and 1926: the largest of these was at Old Oak Common (Fig 8.13). The smaller depots had engine sheds with straight longitudinal roads. The eight larger ones, including Old Oak Common, were based on the old 'roundhouse' layout, with roads arranged radially around turntables.[24]

The company had been building new sidings for goods wagons and carriages at the West London Junction near Kensal Green sidings, and added to them as recently as 1892.[25] However, more was required. In 1898–9 the GWR were planning their new line to High Wycombe: this was to branch off the main line at Old Oak Common, requiring considerable groundworks

Figure 8.12
Offices and workshops for the engine department, c 1853: workshops for smiths and carpenters in the middle; a dormitory and dayroom for the enginemen to the left; and an office for Daniel Gooch to the right, first floor.
[Network Rail/EH AA031083]

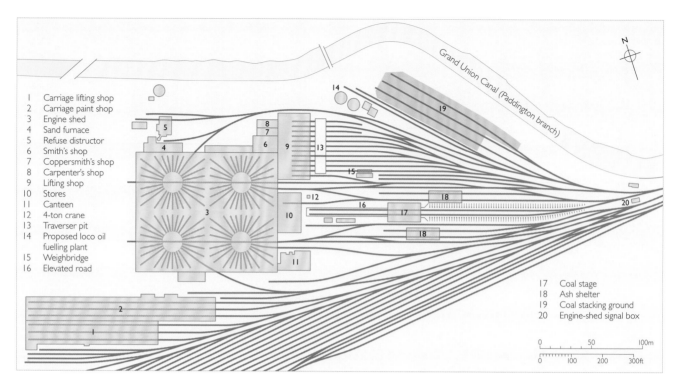

1	Carriage lifting shop
2	Carriage paint shop
3	Engine shed
4	Sand furnace
5	Refuse distructor
6	Smith's shop
7	Coppersmith's shop
8	Carpenter's shop
9	Lifting shop
10	Stores
11	Canteen
12	4-ton crane
13	Traverser pit
14	Proposed loco oil fuelling plant
15	Weighbridge
16	Elevated road

17	Coal stage
18	Ash shelter
19	Coal stacking ground
20	Engine-shed signal box

Grand Union Canal (Paddington branch)

Figure 8.13
A plan of the GWR's
locomotive depot at Old Oak
Common, based on a plan
of 1946.
[© English Heritage]

there. Providentially, there was a large adjacent area of common grazing land available for development. In November 1898 the general manager proposed to move the carriage sheds to Old Oak Common, together with a sidings and a general goods and coal depot. In October 1899 the decision was taken to build new engine sheds there rather than at West London Junction, and in December Churchward presented an estimate of £70,000 for a new engine shed.[26] Work went ahead on the High Wycombe branch from 1900, and the Old Oak Common site seems to have been purchased and cleared. In October 1901 the designs for the engine shed were being modified, but in the event only the carriage sidings were laid out for the time being.[27] The idea of a locomotive depot was revived by Churchward in 1903. In June he presented the board with plans for Old Oak Common with an estimated cost of £110,000. The main contract for the engine shed, factory, stores and ancillary buildings was awarded to Messrs William Walkerdine of Derby, for £40,313: work began in January 1904. By October 1904 Churchward could start ordering the mechanical fittings for it. In 1905 a further contract was awarded for the new carriage shed, completed later that year. The locomotive depot opened in March 1906.[28]

At Old Oak Common the GWR had acquired a large triangular site, bounded to the north by the Grand Union Canal, to the south by the main line (which bends to the south here), and to the west by Old Oak Common Road. The site was divided into three approximately equal triangular segments, their apexes pointing east toward the various junctions with the main line. At the southern edge of the site the two tracks of the High Wycombe line branch off, heading west. North of this, the first big segment of the site was given over to a large area of carriage sidings with 26 tracks and, north of that, a vast new carriage shed housing 15 lines of track beneath a triple-span roof (Fig 8.14). The contract for this was let in 1905, and the drawings are dated 1905–6.[29] The middle segment of the site was initially occupied by a vast marshalling yard with around 35 lines of track.[30] The northernmost segment of the site was given over to Churchward's new locomotive depot.

The locomotive depot was dominated by a vast new engine shed: this was demolished in 1964–5, but in its heyday it would have been an awe-inspiring sight (Fig 8.15). Oblong in plan and measuring 444ft by 360ft (135.3m by 109.7m), in layout it effectively comprised four roundhouses put together: that is, it was divided into quarters laid out around four 65ft (19.8m)-diameter turntables, each with 28 lines radiating from it. The turntables were supplied by Ransomes & Rapier of Ipswich: unusually at this date, they were electrically operated. The engine shed was built of red brick and paved with blue engineering brick. The building was roofed in six conventional double-pitched spans running east–west, with raised top lanterns running longitudinally. The roofs were carried on remarkable 60ft (18.3m) trusses: roof trusses of similar design over the lifting shop and stores survived until 2011 and are described later. Churchward was probably responsible for the overall planning himself, though the detailed design and specifications were doubtless provided by William Armstrong, the GWR's new-works engineer, and his staff. The whole complex including the turntables was powered by electricity from the outset, supplied from the company's generating station at Park Royal. Long inspection pits were built beneath the tracks running into the shed, and past the coal stage.[31]

Abutting the edges of this vast building were a number of subsidiary buildings, which apparently went up at the same time. On the south a small block housed a sand furnace, later converted to a mess room: this survived until 2011. The west side was apparently left clear. To the north there was another sand furnace: an 'economical boiler washing plant' was added in this area in 1910, and a refuse destructor plant in 1927: all this apparently went in 1964–5.[32] Abutting the north-east corner was the largest of these buildings, the 'factory' or lifting shop

with an associated smithy, and abutting the east side to the south of this was a substantial store block: these survived until 2011, and are described later.

Churchward retired as chief mechanical engineer in 1922, and was succeeded by C B Collett. Collett built on Churchward's achievements, designing the famous 'Castle' and 'King' classes which pulled the GWR's expresses in the 1920s and 1930s. A proposal in 1927 to double the engine shed in size to accommodate a further four locomotive turntables was never realised. In 1935 the Government was

Figure 8.14
The vast carriage sheds at Old Oak Common, apparently still bearing the scars of wartime damage, June 1972.
[Great Western Trust Collection]

Figure 8.15
The engine house at Old Oak Common, taken in 1911.
[Great Western Trust Collection]

encouraging large companies to invest in large-scale works, to help relieve unemployment: the Ministry of Transport approached the 'Big Four', offering to underwrite projects. The GWR agreed to fund £5.5 million worth of work around their network: this included £303,000 towards 'rearrangement of Carriage Sidings and a Carriage Repair Shed' at Old Oak Common.[33] These occupied the northern slice of the marshalling yard in the middle segment of the site, and were thus immediately south of the main engine shed. Plans were ready in August 1938, and the completed buildings opened in April 1940: they survived until 2011. Work on ancillary buildings was still going on: a new office for the head foreman was only finished in 1942. Air-raid shelters were installed, and anti-aircraft batteries set up nearby. Despite this, Old Oak Common was attacked by enemy aircraft on a number of occasions.[34]

During the war and after it, coal shortages were leading the 'Big Four' to think about alternatives to coal, and in 1946 the GWR decided to convert 10 coal-burning locomotives to run on oil: oil tanks and fuelling plant were added to Old Oak Common. In the same year the board also authorised the purchase of a proto-type gas-turbine locomotive designed to run on heavy fuel oil, from the Swiss manufacturer Brown Boveri. By the time it arrived in 1948, the company had been nationalised: British Rail went ahead, installing fuel tanks for gas turbine and later diesel locomotives at Old Oak Common c 1949–52.[35] In the last few months of its existence, the GWR approved heavy expenditure on maintaining and modernising Old Oak Common: proof of the company's vitality and faith in the future, up till the end.

After British Rail took over, steam locomotives were gradually replaced by diesel multiple units (DMUs), but for the time being they had to stable and maintain both. The main engine shed was still in use for steam locomotives, and the lifting shop contained facilities that they could not do without, so the carriage paint shop was converted as a repair and maintenance shop for the 'Blue Pullman' DMUs in 1959–60.[36] Steam locomotives were gradually withdrawn from the network in the early 1960s and servicing their diesel replacements did not need such extensive covered accommodation, so Churchward's vast engine shed was now surplus to requirements. Its demolition began in March 1964, and the last steam locomotive left the depot in March 1965. A new servicing shed

with three through-roads was built for diesel locomotives on the south-east part of the 1906 engine shed site: a new turntable was installed behind this, on the site of the 1906 turntable in the south-west quarter of the old shed. The 1906 'factory' or lifting shop and smiths' shop were re-equipped as maintenance space. The remodelled diesel depot was opened by the chairman of the British Railways Board on 20 October 1965, with facilities for maintaining 70 main line and 25 shunting locomotives.[37] The end of steam power brought a similar upheaval all over the network, and it would seem that most of Churchward's depots of 1906–26 were demolished around this time, not just Old Oak Common.

The 'Blue Pullmans' were withdrawn from main-line service in 1973, and their servicing depot in the former carriage paint shop was no longer needed. Instead, a new three-road servicing shed for high-speed trains was built on the sidings to the south in 1976, and extended 10 years later. Old Oak Common remained in constant use, with other more minor changes, until c 2009.[38] The former locomotive depot at Old Oak Common has been chosen as the principal site for stabling and servicing locomotives for the new Crossrail services due to enter service in 2017–18. The site was cleared and all its buildings demolished in 2011–12. A thorough assessment of the standing buildings was carried out in advance of this. There is only space here for a brief description of the buildings as they stood until then.

The demolition of Churchward's vast engine shed in 1964, and its replacement with a much smaller through-shed for diesels left something of a void at the heart of the Old Oak Common depot: some of this was left as waste ground but some was paved in concrete and the south-eastern turntable pit, the one that had been re-formed in 1964–5, was still in place. To the south and east stood sections of the outer walls, with the surviving ancillary buildings abutting them, until 2011.

The most important of these was the locomotive lifting shop, or 'factory' in GWR parlance, which originally abutted the north-east corner of the engine shed. Its main space was 195ft (59.4m) long by 101ft (30.8m) wide: on three sides the outer walls were of load-bearing brickwork. The fourth side was steel framed and originally had 12 continuous entrance doors beneath flat lintels, beneath which 12 lines of track led into it. The building was roofed in two spans of differing height, carried on a central row of steel stanchions. The lifting shop was so-named as its main purpose was to allow for the raising of locomotive bodies to allow access to the wheel bearings beneath, a task commonly performed with jacks or lifting tackle but here carried out with a powerful overhead electric crane, housed beneath the higher of the two spans. The crane had to be carried on a heavy framing, so the building was partly steel framed (Fig 8.16). The central line of stanchions supported fish-bellied beams of riveted steel plate, relatively unusual in building construction, which carried the two roof spans and one side of the crane rails. The original crane was replaced by a 50 ton (50.8 tonne) crane made by S H Heywood & Sons of Reddish in the 1920s whose principal member was also a fish-bellied beam of steel plate: this survived until 2011. At the back of this main space was a narrower single-bay space, also dating from 1905–6 and originally built as the smithy. Subsequently, probably in the British Rail period, it was adapted with a single line of railway track run into it with an inspection pit. To either side of this eight large lifting jacks were mounted on rails, positioned in pairs: they probably dated

from the 1960s, and looked as if they were intended to lift engine casings. Off to one side was a smaller boiler house, also of 1906. In the tracks just east of the lifting shop was an electric traverser to move locomotives from one track to another.[39] A little further east was a big oblong water tower on a brick base (Fig 8.17).

Further south, and also originally abutting the 1906 shed was another brick building with two double-pitched roofs aligned east–west, which originally abutted the east side of the main engine shed: this was originally the main store building. On the south side of the 1906 engine shed was a relatively small and simple brick building, originally erected to house a sand furnace but converted to a mess room in 1937.[40]

The store and the lifting shop both retained their original roof trusses of 1906. These were composite trusses of wood, iron and steel of a remarkable design, certainly by the GWR's own engineers. Brunel had designed several roofs for the company which were carried on trusses of wood trussed with wrought iron, so this might be regarded as a continuation of Brunellian traditions. The principal members were of softwood, with iron caps for their feet

Figure 8.16
The 'factory' or locomotive shop at Old Oak Common with La France, *one of three French locomotives bought by G J Churchward to test his own engines against.*
[Great Western Trust Collection]

Figure 8.17
The coaling station and
water tower at Old Oak
Common where locomotives
were supplied. The gables
in the background are the
storehouses attached to the
main locomotive shed, with
the end gable of the 'factory'
or locomotive lifting shop to
the right.
[Great Western Trust
Collection]

and iron connectors, trussed with rods of steel or wrought iron. The design was effectively a form of Polonceau truss, in which each of the principal members are trussed individually on the underside: the two isosceles triangles thus formed are tied across the mid-span with tie-rods, which link connecting pieces at their apexes (Fig 8.18). In this case, the connecting pieces were elaborate castings of a curious flattened O-shape, made asymmetrical by the seatings for the various timber beam ends. They were almost certainly cast in steel: very few organisations at this date would have had the skills to make steel castings of this size and elaboration, but the GWR, of course, did. The iron and steelwork for the 1905–6 buildings is known to have been supplied by the company, probably from Swindon. The same truss design was used for the main engine shed here, and at other GWR locomotive depots built around the same time.[41]

South of the site of the engine house the immensely long carriage shops, built under Collett's direction in 1938–40, were standing until 2011. These comprised two parallel ranges, similar in width and overall design but of unequal height and length, with brick outer walls carrying steel-framed roofs. On the near (north) side was the longer and lower range,

built as the carriage paint shop but later adapted for maintaining the 'Blue Pullman' diesel units. This was 592ft (180.4m) long and 69ft 6in. (21.2m) wide, its roof carried on simple trusses of steel angles with continuous glazed lanterns: it had extensive ventilation apparatus to cope with the painting operations. Adjoining this to the south was the carriage lifting shop, 412ft (125.6m) long and 70ft 5in. (21.5m) wide. This was higher and had a steel frame to accommodate a 20 ton-overhead travelling crane, which ran the whole length of the building.[42] These ranges were not as structurally remarkable as the 1906 lifting shop, but were of historic interest for the way they were designed around their functions, as elements of one of the GWR's most important depots (Fig 8.19).

The locomotive lifting shop was undoubtedly the most important and interesting surviving element of the site. Until its demolition, this was the last surviving workshop building directly associated with G J Churchward, an outstanding figure in the history of the GWR, and of locomotive design generally: almost all the depot and workshop buildings that Churchward caused to be built c 1906–26 were demolished when steam haulage came to an end in the 1960s. The lifting shop's combination of traditional load-bearing brickwork,

Figure 8.18 (above)
The remarkable roof frame
of the 'factory' building
had composite trusses of
timber and wrought iron:
the oddly shaped connecting
elements in the middle were
probably steel castings made
in the GWR's foundries at
Swindon.
[© Crossrail Ltd]

Figure 8.19 (left)
The carriage lifting shop at
Old Oak Common, with one
of its two original overhead
travelling cranes still in situ,
2010.
[© Crossrail Ltd]

with the modern materials and technology of its steel framing and electric crane, and the individual touch of the Brunellian Poloceau trusses, made it very characteristic of the GWR's history and engineering culture.[43]

Carriage sheds, wagon sheds and stores

From an early stage, the GWR ran into difficulties caused by shortage of space and by the narrow 'throat' of their trackbed as it ran into the terminus. For the carriage department, Brunel planned a long series of buildings running along the south side of the tracks. Like the goods station and the locomotive department, their appearance could be reconstructed in detail from the sets of contract drawings in Network Rail's archives. From the Bishop's Road Bridge to the Westbourne Bridge was a continuous series of buildings, a short range of offices and stores followed (going west) by sets of carriage sheds one and two storeys high and with an overall length of around 250ft (76m).[44] Just beyond the Westbourne Bridge was an irregularly shaped building housing the department's stores and offices.[45] Beyond this was a large group of workshops, including an L-shaped group of smiths' shops, one wing of which was 107ft (32.6m) long and the other about 120ft (36.6m).[46] Further west was a huge wagon shop, 275ft (83.8m) long and

94ft (28.7m) wide, for which another set of contract drawings survive (Fig 8.20).[47] All of these buildings seem to have been swept away after 1900 to make room for an extended milk platform between the Bishop's Road Bridge and the Westbourne Bridge; the area was again redeveloped in the 1930s as the main down parcels depot. In 1905–6 a vast new carriage shed and sidings were provided at Old Oak Common, as noted earlier.

Former GWR stationery store, Porchester Road

Some way to the west of the station, on the south side of the tracks next to Lord Hill's Bridge, the stationery store is a square block of four storeys over a track-level basement. It was built in 1906–7 to rehouse the stationery department, previously in the arrival side of the terminus, and was thus one of the preparatory works for the remodelling of the station between 1907 and 1914. The stationery store was designed by William Armstrong, the GWR's new-works engineer, and its whole structure, including the floors, represent one of the company's earliest uses of the Hennebique reinforced concrete system. The building has handsome Classical detailing, of hard red brick with dressings of patent Victoria stone. It incorporated an electricity substation, supplying power from the GWR's new generating station at Park Royal

Figure 8.20
An end elevation for the new carriage sheds, c 1855. [Network Rail/EH AA042117]

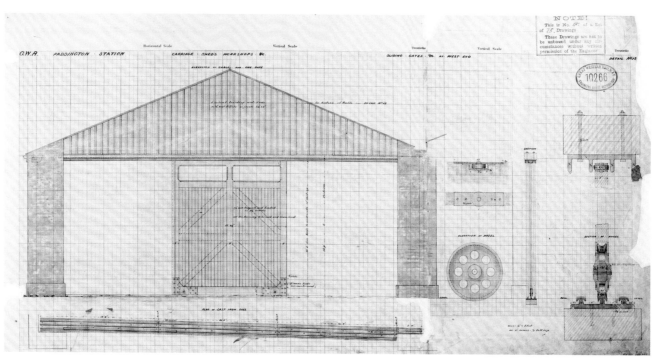

both to Paddington and to the Metropolitan Railway. The building is currently being refurbished for office use by Network Rail.[48]

Former down parcels depot, 4–14 Bishop's Road

After the construction of the GWR's main line, a restricted triangular site was left over between the tracks to the north, the Bishop's Road to the south and Westbourne Terrace to the south-east. The company's long carriage sheds rose on the north side of this and the rest of the site was filled with tall terraced houses, and Bishop's Mews behind. The GWR bought up the leases and replaced most of the houses with two large concrete-framed buildings as part of its 1930s expansion.[49] The parcels depot, designed by P A Culverhouse and built between 1932 and 1933, occupies the eastern part of this site and looks south onto the Bishop's Road. It is of five storeys plus a mansard attic, with a simple Classical façade of brick and patent stone.[50] A low wing of the building facing over the tracks housed the departure side signal box. The building has long since been sold, but behind the façade the depot yard and platforms remain under their plain steel-framed roof, in use as a National Car Park. From platform 1 a rewarding walk may be had, right through the depot and beyond as far as the Ranelagh Bridge. The long timber cover in the depot platform represents the duct which took electrical cables from the substation under the Porchester Road stationery store to the passenger station.

Former GWR stationery store and hostel for GWR employees, 167–9 Westbourne Terrace

From the outset the GWR found that in order to run daily train services they had to be able to accommodate train crews overnight at their principal stations. Around 1852 a staff hostel was built just outside the 'country end' of the new Paddington train shed. Fitted between the tracks, it was a rectangular building in a simple 'engineers' classical' style, 10 bays long and 2 storeys high.[51] A couple of plans and historic photographs record its appearance and it is just visible in the background of Frith's painting *The Railway Station* (*see* Fig 4.2). It was demolished in *c* 1906 to make way for the major redevelopment of the station.

Between 1932 and 1935 a new staff hostel was built to designs by P A Culverhouse, abutting the parcels depot, south of the tracks and hard by the Westbourne Bridge (Fig 8.21). It is concrete framed and of six storeys, with façades towards the road and the tracks in the same restrained Art Deco style that Culverhouse employed on the arrival side offices, with similar shallow oriel windows. The top two floors were a hostel for women staff, mostly employed in Paddington's refreshment rooms, and Culverhouse provided them with a spacious rooftop terrace (Fig 8.22).[52]

Figure 8.21
The stationery stores and staff hostel under construction, c 1934.
[Network Rail]

Figure 8.22
Workers from the station restaurant playing table tennis on the roof of their staff hostel, 19 August 1935
[Getty Images]

The former Mint stables, Winsland Street and South Wharf Road

The Mint stables, now the Mint Wing of St Mary's Hospital, Paddington, are a rare historical survival and a vivid reminder that the railways, so far from displacing horse-drawn transport, probably increased the demand for it. The GWR, like all major railway companies, used horses for general road haulage, for parcel delivery services and for shunting wagons and locomotives. In the early days, railway horses tended to be given makeshift accommodation, under railway arches or in odd corners of their goods yards, and most of the companies relied to some degree on livery stables. As the goods traffic grew, these makeshift arrangements were superseded. By the 1850s, better quality purpose-built stables began to appear, such as the LNWR's at Chalk Farm, Camden, which also survive.[53]

In the mid-1870s, the GWR decided that it would be more economic to stop using livery stables and build up their own fleet of horses. By 1890 they had about 1,110 horses working in London out of a total of around 6,000 railway horses in the city. An irregular triangular site north of the station, was available. The GWR already had makeshift timber stables

at the south end of the site, the rest of which was occupied by the Mint public house (from which the stables took their name) facing South Wharf Road, cottages and an open yard. A lease was acquired in 1876 and the company's architect, Lancaster Owen, produced a design for the whole site (Fig 8.23). A contract was signed for a first phase filling the open, northern part of the site. This provided stabling for about 220 horses. In a second phase of work in 1882–3 the southern part was added, leaving the pub surrounded by stable buildings on three sides.[54] This took the Mint stables' capacity to around 500; another 140 animals were kept in a stable adjoining the goods depot. The GWR had a number of out-stations, the largest at Goswell Road on the boundary of the City of London which housed another 200 horses. A 'horse hospital' was established at West Ealing for convalescent animals.[55]

As designed the buildings were two storeys high, of stock brick (Fig 8.24). They looked inwards onto an irregularly shaped yard in the middle of which were separate buildings housing the provender store, mess room and a farrier's shop. Ramps led up to a gallery running around the courtyard to give access to the upper storey. Both floors were divided into rooms housing 10 to 12 stalls each. The floors were of granite setts below and asphalt above,

Figure 8.23
Lancaster Owen's ground-floor plan for the Mint stables, 1876.
[Network Rail/EH AA031059]

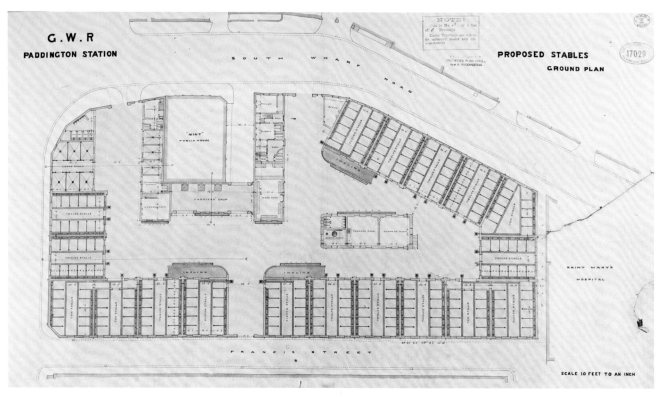

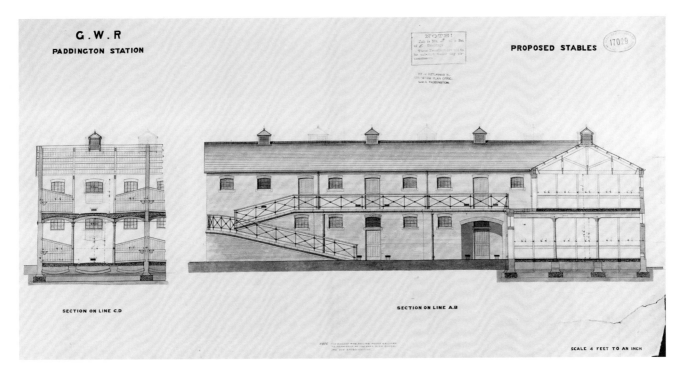

the first floor being laid over brick jack arches carried on cast-iron columns. The roof was carried on king-post trusses. The stalls had wooden dividers and the horses were provided with iron 'Swindon pattern' troughs. In the later part of the building the stalls were separated with swinging bails to guard against horses kicking their neighbours.

In 1910 a major remodelling of the stables was carried out, part of the general campaign of improvements to the station. First, the southern range (that towards the station) was demolished; the site was excavated out and a massive basement of Hennebique ferroconcrete created at platform level as the new up parcels depot. Lifts took the parcels up to a circulating area with six loading docks for the GWR's horse wagons, opening onto London Street and Francis Street.[56]

As the south range had been given to the parcels depot, an extra storey was added to the whole Mint stables group to make up for the lost space. Extra columns were inserted into the existing buildings, some of them improvised by placing two lengths of rail back to back. The work was carried out one 60ft (18m) section at a time so the stables could be kept in use. The GWR, ingenious and economical as ever, literally raised the roof by removing the slates over a section, inserting screw jacks at 20ft (6m) centres and lifting the roof frame as the walls were carried up. The original king-post

trusses are still in place. Outside, the galleries and ramps were rebuilt to give access to the first and second floors: like the new basement, they were of Hennebique ferroconcrete (Fig 8.25).[57]

The remodelled stables housed 50 horses in the basement, 261 on the ground floor, 288 on the first floor and 66 on the top floor (Fig 8.26). The historian Robert Thorne has likened the remodelled courtyard to 'an open-air theatre, set to life twice a day when the horses were led to and from their duties'.[58] The working week started at 2 am on Monday morning and the horses might be out for up to 18 hours. A heavy

Figure 8.24
The Mint stables – a cross section and part-elevation, 1876.
[Network Rail/EH AA031060]

Figure 8.25
The Mint stables, horses being led down the ramps, early 20th century.
[Great Western Trust Collection]

Figure 8.26
Horses in the Mint stables,
1936.
[© National Railway
Museum/Science and
Society Picture Library]

railway van might weigh 2 tons, and carry another 7, requiring four horses to pull it, a total mass of 13 tons (13.2 tonnes) moving at a slow walk (*see* Fig 5.14). A light parcel cart would be drawn by a single horse, trotting at around 8 miles (12.9km) an hour.[59]

The stables housed horses into the 1950s. Thereafter, British Rail converted them for a variety of other uses: the upper floors became the chief engineer's drawing office for a while. It was probably about this time that the ramps and galleries were glazed in and the second-floor windows enlarged. Most of the stable fittings, of course, were lost. In 1965 the stables were acquired by St Mary's Hospital and further alterations carried out to fit them for new purposes. They are listed at Grade II.

Motor vehicle maintenance depot, now offices, Harrow Road

In 1967–9, Paddington's estate was enhanced with a building of outstanding quality, though it may seem paradoxical that it was built for road vehicles rather than railway rolling stock (Fig 8.27). In the 1960s, the construction of the Westway flyover transformed the setting of the station, vividly expressing the new dominance of road transport. British Rail, like the GWR,

maintained a vast freight operation. Just as the GWR had needed hundreds of horses, so British Rail found that it needed large fleets of road vehicles: the depot was the mid-20th-century equivalent of the Mint stables.[60]

A cramped, interstitial site was available, hemmed in between the canal, the Harrow Road and the Westway, at the point where a slip road drops from the Westway to a roundabout. The site was blighted by the flyover, lacked a conventional street frontage and over 80,000cu yd (61,000cu m) of earth had to be extracted before work could begin. These drawbacks were brilliantly exploited by Paul Hamilton, who designed the depot in 1967. The east block, within the well of the roundabout and housing the garage and the maintenance area, is a long oval, with a raked leaded roof with oval lights and central top-lighting.

The west block occupies a difficult triangular site hemmed in between roads. It is on five levels, rising to seven at its east end. At the bottom, at the level of the maintenance area, was plant and storage, and above this workshop space. On the third floor there was a machine shop and on the fourth and upper levels, mess rooms and offices. The upper floors are slightly jettied: this, the curved outline and the continuous bands of glazing, give the building an elegant 'moderne' quality. The paired chimneys

Figure 8.27
The former motor vehicle
maintenance depot seen at
night.
[© English Heritage
K010969]

rising at the western 'prow' of the building enhance its rather nautical quality. Inside, the staircase, dramatically curved and top-lit, echoes the shape of the building. The maintenance depot reads to great advantage from the Westway, from where its curved, jettied forms seem to advance and recede as one approaches and moves past, a rare instance of architecture relating positively and successfully to modern highway engineering. The building was sold after British Rail's privatisation, but has been very successfully renovated as office space. The journal *Official Architecture and Planning*, reviewing the newly completed building in

February 1969, remarked that 'if British Rail continue to appoint talent of this quality, in AD2068 a preservation society will nominate the Paddington Maintenance Depot one of the most significant railway buildings of its era and in its urban context'.[61] It is satisfying to record that the depot's outstanding quality was recognised rather sooner than that, and that it was listed at Grade II* in 2001.

Bridges over the main line

When the GWR widened and replanned their main lines between 1903 and 1911, they were obliged to rebuild 10 bridges. Originally built as brick viaducts, these were all replaced with long-span steel girders, designed by the GWR's own engineering department. The first four of these are considered here, starting with the furthest away and working back towards the station.

Lord Hill's Bridge

This, the fourth bridge out from the station, was named for General Sir Rowland Hill, 1st Viscount Hill, one of Wellington's most liked and trusted commanders. He was created Commander in Chief in 1828 and lived at Westbourne Place, a substantial villa just to the south of the GWR's main line. The original

viaduct had five brick arches, two of them later replaced with cast-iron girders. The brief for rebuilding was complicated by requirements that the bridge should climb gently from south to north and carry gas and water mains. It also had a very tight clearance: to maintain a minimum height of 14ft 6in. (4.42m) over the tracks, the road deck of the south span could only be 3ft (0.9m) deep. The Horseley Bridge Company made the very heavy hog-back girders for the two spans, 1,112 tons (1,130 tonnes) of steel being used in all.[62]

Ranelagh Bridge

Grissell & Peto signed a contract to build the original nine-arch brick viaduct on 2 October 1837.[63] It was replaced with two spans of N-braced hog-back girders, the southern one of 160ft (49m) across the main lines, the northern one of 140ft (43m) over sidings, completed in the summer of 1909 (Fig 8.28). Much care was taken with the detailing of the impressive granite piers and the decorative mouldings at street level.[64]

Westbourne Bridge

The original viaduct, for which Grissell & Peto signed a contract on 26 June 1837, had 10 brick arches.[65] At an unknown date, 9 of the

Figure 8.28
The Ranelagh Bridge, before and after rebuilding. Note the ticket office for the Royal Oak underground station, in both views. The station platform is just visible beyond the bridge.
[Great Western Trust Collection]

10 spans were replaced with flat iron girders. The bridge was rebuilt in *c* 1907–10, the contract for £37,562 being let to Messrs Andrew Handyside of Derby on 6 August 1907 and the bridge opened in 1910.[66] It crosses the line at a skewed angle and is thus much longer than the other bridges. The two spans rest on impressive granite piers and are carried by exceptionally long bowstring trusses: the southern span has two trusses each of 232ft (70.7m), the northern span has trusses of 215ft (65.5m) and 244ft (74.4m), of differing lengths because of the skewed plan. The steelwork, made by Handyside, weighed 2,150 tons (2,184 tonnes). The sides of the roadway are enclosed with solid iron panels and at either end there are big Classical pedestals, also of iron. The bridge is listed at Grade II.[67]

The Bishop's Road Bridge, and the Paddington LTVA (Long Term Vehicle Access) Project

When the GWR obtained the Act of Parliament for its extension to Paddington in July 1837, one of the major conditions attached to it related to the construction of a number of bridges. The easternmost of these carried a new road, the Bishop's Road, from Westbourne Place to join the Harrow Road. This involved spanning the shallow valley in which Brunel was putting the tracks and station with a long brick viaduct. The contract for this was let in 1837 to Messrs B and N Sherwood. Its 26 arches must have gone up very quickly, for the temporary terminus went up around them and was (more or less) ready for opening in June 1838. The original viaduct is described in chapter 1.

Brunel's brick viaduct seems to have remained largely as built until 1906, when the greater part of it was demolished to allow for realignment of the railway tracks into the station. The GWR board approved a vote of £24,000 for the purpose on 14 June 1906, and the main contract (for £17,843) was let to Messrs Jackaman on 19 July.[68] They replaced the viaduct with a 200ft (70m) single-span hog-back truss of riveted steel, designed by the GWR's chief engineer William Armstrong and manufactured by Westwood & Company, at Millwall on the Isle of Dogs (Fig 8.29).[69] Like its predecessor it only carried a two-lane highway, and by the late 20th century it had become a notorious traffic bottleneck. Furthermore, long-term plans had been developed by Westminster City Council and Network Rail to improve vehicle access to the station by making the main route in for cars and taxis from the west side via a new and wider bridge, the Paddington LTVA (Long Term Vehicle Access) project: the background to this is summarised in chapter 6. After a long campaign of planning

Figure 8.29
The Edwardian Bishop's Road Bridge in 2003 before its replacement.
[© English Heritage K010993]

the 1906 Bishop's Road Bridge was demolished and replaced with a modern five- to six-lane bridge carried on giant steel beams in 2004–6.

This bald statement masks a very complex project: the replacement of a large steel bridge with a new and wider bridge over one of the busiest stretches of railway line in Britain and with several other planning obstacles involved presented a major challenge. This was addressed by a 'bridge launch', the most ambitious use of this technique yet seen in Britain. First, the northern section of the bridge was demolished: this included six of the original brick arches, much refaced, and the section spanning the canal (which presented a whole separate issue, discussed later). The northern section of the new bridge was built on this site in 2004, and a large concrete supporting pier immediately north of the tracks. Then, scaffold towers were erected at the corners of the 1906 steel span, with cradles from which it was hung: the whole span, 60m long and weighing 941 tonnes, was jacked high into the air by a technique called 'strand jacking' in the course of a night in August 2004, while the railway beneath was closed. There it remained for over a year, while its replacement was constructed on the already built northern section. In 2005 the new bridge, 105m long and with a launch weight of 2,500 tonnes, was launched diving-board fashion between the scaffold towers carrying the 1906 span in its raised position (Fig 8.30). The launch was carried out gradually, in 21 stages, sliding over the new pier and across the railway lines until eventually it reached the far (Eastbourne Terrace) side. Then the 1906 span was jacked back down onto the new bridge-deck and broken up there. The new Bishop's Road Bridge opened to traffic in June 2006: there can be no doubt that Brunel would have been mightily impressed.[70]

The Paddington LTVA project cost around £65 million, its cost shared between several partners. It is curious to reflect that replacing Brunel's brick viaduct with the 1906 bridge had cost the GWR £24,000: and that in 1906–10 they rebuilt the next three viaducts down the line in similar fashion, as well. We do not know how Armstrong and Grierson managed all this: given the formidable complexity of the LTVA project, it would be interesting to find out.

The Paddington LTVA project was also affected by another entirely unforeseen event, the discovery that the bridge's canal spans represented Brunel's earliest surviving iron bridge, and were a unique piece of engineering history.

The Bishop's Road Canal Bridge

In designing his new Bishop's Road Bridge, Brunel had to cross the arm of the Grand Junction Canal which led to Paddington Basin. The canal represented both a problem and an opportunity for the GWR. The benefits of having their terminus next to it were clear enough, and it provided a ready means of bringing materials to this end of the line, but at the same time it presented him with a problem with levels. The main part of the new bridge over the railway tracks did not present a problem: he had ample clearance and could build conventional brick arches. The canal, however, was higher up, almost at street level. The new Bishop's Road Bridge would have to clear the canal, giving the Grand Junction Canal Company a respectably wide span for their barges and a minimum clearance of 10ft (3m) above the water and towpath, before dropping down to meet the Harrow Road. On this downhill section Brunel was obliged to provide a fairly gentle gradient of not more than about 1:20. This added up to a tight problem of clearance and headroom: to solve it he had to minimise the structural depth of the canal bridge, so a conventional brick arch was out of the question.[71]

Figure 8.30
The Bishop's Road Bridge, temporarily elevated in 2005, with its replacement being gradually 'launched' beneath it: a major engineering operation carried out over one of the busiest railway lines in Britain.
[Dorman Long Technology]

Railway engineers faced situations like this, where their line had to cross or go under a road or a canal with tight headroom, or at a difficult angle, or both – on many occasions. Brunel encountered such problems for the first time in three places near the London end of the GWR's main line, and he addressed them by designing cast-iron bridges. Two were skewed crossings over canals: one on the main Grand Junction Canal at Southall, and another (very close to it) over the Paddington Branch of the Grand Junction Canal at Hayes. The third bridge was much more complex, as it spanned a crossroads where the Metropolitan Turnpike Road from London to Uxbridge, the present A4020, intersects with Windmill Lane and Greenford Lane. The railway was to cross the turnpike road at an extreme skew of 67 degrees: Brunel persuaded the turnpike commissioners to accept a slight kink in their road which reduced the skew to a just manageable 60 degrees, and set to work on his design.

He was relying for his understanding of cast iron on the theoretical work of a London-based scientific writer, Thomas Tredgold, and seems to have been unaware of the more up-to-date and accurate work of a Manchester-based scientist, Eaton Hodgkinson, who was assisting Robert Stephenson with the design of cast-iron bridges on the rival London & Birmingham Railway at the time. The central section of the Uxbridge Road Bridge over the crossroads was to be carried by two massive I-section beams with equal top-and-bottom flanges, as suggested by Tredgold. They were 4½in. (114.3mm) thick (in iron casting terms, very thick indeed), with secondary beams spanning between them to carry the railway. The secondary beams were fixed into the primary beams with mortice-and-tenon joints, essentially a carpenter's detail. The design was seriously flawed: Brunel did not realise that when casting an iron beam, the thicker it is, the more likely it is to develop flaws. The box-outs for the mortices could have impeded the flow of molten metal, adding further to the risk of flaws. The upshot was that in June 1837 one of the two primary beams broke: the railway was not yet open, but the bridge would have been carrying spoil-trains for building the huge embankments to either side of it. It was rebuilt with a new beam cast to the same design, in time for the opening of the line in May 1838.[72] However, the fiasco had alerted Brunel to the potential problems of using cast iron, and

certainly influenced the design of his next iron bridge – the Bishop's Road Canal Bridge. This was designed in the spring of 1838.[73]

On 18 May 1838 he wrote to R C Sale at the canal company's office, as follows, (displaying his characteristic economy with punctuation):

My dear Sir,

I forward you an elevation of the Bridge we propose to erect over the basin at Paddington in lieu of the present foot bridge by the provision of our Act it is to be a carriage road bridge – 40 feet [12.2m] wide and to communicate with the Harrow road which limits out height. The bridge will therefore be of Cast Iron.

I have made the principal opening as large as I possibly could consistent with safety and have provided a second Arch which is also large enough for barges to pass freely. And the whole is as wide as the present basin.

I propose to get in the foundations during the stoppage of your canal next month and shall therefore feel obliged by an early reply. If Mr Holland wishes to see me on the subject, I shall be happy to meet him, but I must ask him to call on me, as I am still somewhat of an invalid. I am dear Sir,
Yours very truly, I. K. Brunel.[74]

Mr Holland, who seems to have been the canal company's engineer, evidently objected. On 25 May Brunel wrote to Sale again:

My Dear Sir,

I think Mr Holland does not bear in mind that the basin and Towing path at the point of crossing is upwards of 60ft [18.3m] with the head way we have it would be impossible to build such in one opening, perhaps a larger opening than the one I have proposed might be made although really I should not like to try it, but there would not be room to navigate the small arch – by the present proportion.

The large opening is very ample probably the largest on your Canal while the side arch is still quite large enough for all barges – moving about without a tow rope.

With respect to the loss of water, that difficulty also may be removed as it is proposed to drive a stank or dam across the basin and thus save the water south of it, and to drain only that part of it between the stank and the present bridge.

I trust these explanations will be quite satisfactory and that so far as the sanction of your company is necessary we may proceed. A plan shall be sent.
I am my dear Sir,
Yours very truly, I.K. Brunel.[75]

On 30 May the directors of the Grand Junction Canal Company considered the design and accepted it, a little reluctantly.[76]

For all the difficulties of the job, Brunel had produced a very handsome tripartite design, with brick piers, a Portland stone cornice and a fine cast-iron railing. The middle and northern spans had very shallow segmental arches formed of cast-iron beams or girders carrying cast-iron plates. The southern span, over the towpath, was a simple brick arch, but was given a facing to match the northern span.

The main contract for the canal bridge was let to Messrs Sherwood on 27 October 1838. The total price has not been found, but Brunel recorded the tender prices in one of his notebooks and it shows that the Sherwoods offered to provide the cast iron for 'ribs, bedplates, bridging pieces, covering plates, railings …' at £15 12s 6d per ton.[77] They subcontracted the manufacture of the ironwork to Messrs Gordon of Deptford.

Brunel had learnt to be cautious about using cast iron from the beam failure at the Uxbridge Road Bridge, and there is interesting evidence of how he load-tested the beams for his new bridge in one of his volumes of 'Facts', a series of notebooks in which he entered research notes and information of all kinds.[78] Four pages record the results of load-tests, dated December 1838 and supervised by J Colthurst, one of his assistant engineers. A drawing shows the beam designs and how they were tested: the arched beams were placed, toe to toe, in a heavy wrought-iron frame, and pressure was applied with a hydraulic press. (Fig 8.31).[79]

Every beam to be used in the bridge was tested. Those for the main span, which were 35ft (10.7m) long, were pressed or squeezed by the hydraulic press to 20, 25 and 30 tons (20.3, 25.4 and 30.5 tonnes) of pressure. Brunel's assistants measured the deflection – the degree

to which the beams were bent out of shape, recovering their original form afterwards. Cast iron is a very stiff material, and at 30 tons the beams were only bending by ⅝in. (16mm) to an inch. One of them broke at 28 tons strain, showing that the beams were being tested to the limits of their strength. Brunel drew it in cross section, noting that 'the parts marked with dots composed of nothing but Slag, refuse cinder & Sand'. The smaller girders for the 16ft (4.9m) arch were also tested to 30 tons: these deflected by around ¾in. (18mm). The outermost beams, which would have to carry the parapets, were thicker, and these tended to deflect by less.

Brunel was thinking hard about the problems of using cast iron, and carrying out other investigations: the same volume of 'Facts' records a series of tests of iron castings from 49 different named foundries, to examine their elasticity, deflection, breaking weight, specific gravity and power to resist impact.

The Bishop's Road Canal Bridge was a unique design, its beams quite different from the I-sections which Brunel had used at his three previous bridges. One might have thought that he would have now followed Eaton Hodgkinson's thinking. Broadly speaking, this argued that the lower flange of a cast-iron beam should be larger in section than its upper flange, as the material is inherently strong in compression but relatively weak in tension. This is because the lower flange or part of any beam under loading is placed in tension (like being pulled from either end) – while the upper flange or part is in compression (like being squeezed from either end). The line within the beam where the two forces are in balance is termed the neutral axis. To Hodgkinson it therefore made sense to give the lower flange more substance, the better to distribute the tensile forces. So he designed beams for Robert Stephenson which were relatively thin in cross section (presumably to address the casting issues), with broad lower flanges and narrow upper flanges like an inverted 'T', and these designs were successful.

Instead, Brunel characteristically went his own way. He designed beams with an elegant arched shape: in cross section, they have a fairly broad lower flange with sloping 'shoulders' and an equally broad bulb-shaped top flange. The upper flanges are actually bigger in section than the lower, so Brunel was heading in the opposite direction to Eaton Hodgkinson here.

No-one had ever designed beams like this before. One might think, given their arched

Figure 8.31
Brunel's sketch design for load-testing the cast-iron girders for the Bishop's Road Canal Bridge, December 1838. The arched beams were to be held in pairs, toe to toe, in a wrought-iron frame, and squeezed by a hydraulic press: the feed pipe for the press can be seen leading off to the left in this image. Brunel recorded the deflection of the beams at 20, 25 and 30 tons (20.3, 25.4 and 30.5 tonnes) pressure.
[The National Archives, RAIL 1149/9 p96]

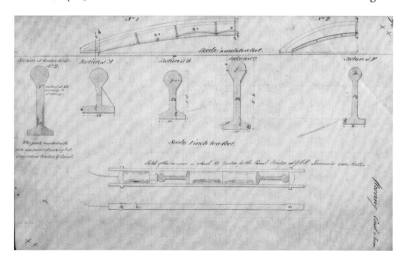

shape, that Brunel was intending them to act as arches, but he didn't give them the kind of 'skew-back' seatings which would absorb arching forces: in structural terms, they were probably acting as beams, not arches. The 'streamlined' section seems more readily explicable: Brunel is likely to have been thinking about the way that molten iron flows into the mould, and trying to produce a streamlined design that would ease the flow, reducing the risk of casting flaws.

The beams supported an elaborate system of cast-iron soffit plates. For the main span these were carried on intermediate 'transverse' members, which sat on the lower flanges of the main beams. For the side span, the soffit plates sat directly on the main beams. The soffit plates, too, were curved. The bridge components must have presented a challenge to Messrs Gordon with its curved lines and complex shapes: their pattern-makers would have had to make timber models or 'patterns' of every one.

The beauty of his design was the ease with which it could all be fitted together. First the brick abutments and piers went up. Heavy cast-iron bedding plates were laid at the arch-springing level. The beams were lowered onto them: there were no holding-down bolts, just some iron cement to act as an adhesive. Then the transverse members were dropped in: again, there were no bolts, but the beams had slightly raised sections to receive them: the transverse pieces were dropped in and secured with wedges. Then the soffit plates were placed in position, again without bolts. Finally, tie-rods

were fixed from beam to beam, with threaded ends and nuts. It not quite clear why Brunel included them: maybe he wanted them to keep the assembly stable during construction. At any rate, the nuts fixing the tie-rods are the only 'positive' fixings in the whole structure. When the ironwork was complete, the area above was filled with a weak lime concrete, and topped off with the road paving.

For some reason Brunel introduced an irregularity into his design: the outermost 'bays' of the two spans were formed, not with the soffit plates he had designed with such care, but with brick jack arches. We do not know why: maybe he did not trust his iron soffit plates to support the weight of the stone cornice and railings. This does seem a little perverse, though, for the jack arches would have added another element of complexity to the construction, requiring Sherwoods to build scaffolding off the canal bed, which the rest of the design would not, in fact, have needed.[80]

Brunel gave his bridge a handsome architectural treatment, with its broad middle arch framed by two narrower arches to either side: to the east was the narrower 16ft (4.9m) iron span, while to the west there was a brick arch of similar width. A fine cornice of Portland stone and handsome railings crowned the composition (Fig 8.32). We can say for certain that this was Brunel's own design, for in addition to the letters quoted above, his original sketch designs for it have been found, corresponding closely to the structure as built.[81] Copies of the contract

Figure 8.32
Brunel designed this cast-iron bridge, probably the first he had built, in 1838 to carry the Bishop's Road over the Grand Junction Canal. Only recently discovered, it has been dismantled to make way for a new bridge and its reconstruction is planned.
[© English Heritage K030904]

drawings have also been found, showing the original design with its handsome cast-iron railings.[82]

On 18 March 1839, the other principal beam on the Uxbridge Road Bridge failed, this time with the railway in use. Brunel rebuilt it using the same beam design again, but was obliged to lighten the bridge deck by replacing it in timber.[83] By this time, he was fully aware of the problems of the material. He designed many more cast-iron bridges, but they were cautious designs, none with spans of more than 34ft (10.4m). He changed his beam designs several times rather than tamely (or reasonably) adopting Hodgkinson's thinking. He carried out his own research and made own designs.[84]

The Bishop's Road Canal Bridge is a unique design representing a specific phase in the development of Brunel's thinking, and is characteristic of his originality of mind. His caution regarding the material was justified, as was demonstrated in 1846 when his friend Robert Stephenson's railway bridge over the River Dee at Chester, which carried on massive cast-iron beams, collapsed while a train was crossing it with the loss of six lives. Brunel gave evidence in support of his friend at the subsequent enquiry.

The canal bridge was such an inconspicuous thing that it sank from notice, and was entirely forgotten as an early work by Brunel. It probably had a narrow escape in 1906–7 when the major part of the Bishop's Road Bridge was replaced with a steel hog-back truss, as discussed earlier. The GWR's engineers, Grierson & Armstrong, presumably inspected it and gave it a clean bill of health. However, at the same time they partly refaced it and removed its handsome railings, replacing them with ugly, high brick parapets. The parapets rendered the canal invisible from the road, and later on the towpath became inaccessible to the public. So the canal bridge sank from notice, seen only by a handful of canal users, until 2003.

In that year, the plans for the Paddington Long Term Vehicle Access (LTVA) project to replace the Bishop's Road Bridge came to a head after over 10 years' preparation. The whole bridge was inspected, judged to be mainly of 1906 and without historic interest, and slated for demolition. The present author, in researching the first edition of this book, came across Brunel's notebooks of 'Facts', with their reference to the 'manner in which the canal bridge for the GWR terminus were tested'. This could only refer to the canal bridge at Paddington. Initial attempts to inspect the bridge were unsuccessful, for the reasons indicated. Westminster City Council was contacted, very shortly before the main contract for the LTVA project was due to be signed. In a meeting shortly after it emerged that here was, indeed, a hitherto unknown iron bridge by Brunel, the only significant element of the original 1830s terminus to remain, but that it was due to be demolished in nine months' time.

Having made the discovery, we negotiated the salvage of the bridge with Westminster City Council, their consultants and contractors. A detailed survey was made by Malcolm Tucker. As part of this, trial pits were dug in the roadway to examine the bridge's construction from the upper side (as it could not be properly understood from the underside) and assess whether it could be dismantled. A photogrammetric survey was carried out by English Heritage's Metric Survey Team. Cass Hayward, design engineers for the LTVA project, produced two alternative salvage schemes. The first, for underpinning the bridge and sliding it sideways, was examined, costed and rejected. The second scheme, for dismantling and removing the bridge, was agreed in principle. By October 2003 funding for this had been agreed by the LTVA partners, while English Heritage agreed to store the dismantled bridge components at Fort Cumberland near Portsmouth.

The Bishop's Road Bridge closed to the public as scheduled in January 2004. The dismantling of the canal bridge went ahead: part-way through, when its iron structure was fully visible, a press announcement of the discovery was made (Fig 8.33).[85] The soffit plates

Figure 8.33
The Bishop's Road Canal Bridge during dismantling.
[© Steven Brindle]

were taken out from alternate bays, and steel lifting-frames built around the remaining bays. The climax of the dismantling came on 31 March 2004 when a giant mobile crane lifted the main bridge-sections out and onto lorries for transport to Fort Cumberland (Fig 8.34). The bridge was cleared by May 2004, without a single day's delay arising to the LTVA project.

The bridge's components, including the whole of its ironwork, the stone cornices and around 15,000 salvaged bricks, are currently in store at Fort Cumberland. Detailed plans for its reconstruction on a site about 183m up the canal have been agreed between Westminster City Council, British Waterways (who will own the rebuilt bridge) and English Heritage. The rebuilt bridge will fulfil a need for a footbridge to link the Maida Vale and Harrow Road areas to Paddington and Sheldon Square. It will also house a café in the northern abutment. The plans are complete but the funding for it is not yet all in place, and for the time being the bridge components remain in store.

The Bishop's Road Canal Bridge is Brunel's earliest surviving iron bridge, and is the only surviving element of his original terminus, the first Paddington Station. Brunel's new station has evolved to meet the challenges of 150 years of convulsive change: it remains one of the busiest and best-loved of Britain's railway stations. The current Span Four and Crossrail projects represent the biggest generational change that Paddington and its neighbourhood have faced since the 1930s. The Bishop's Road Bridge and the iron canal bridge were swept away by the imperative demands of those changes, but there are now good grounds to hope that the canal bridge will soon return to the area, and form the last surviving trace at Paddington of that great moment in the 1830s, when Brunel set out to build 'the finest work in England' and chose this place to be its starting point.

Figure 8.34
A giant mobile crane lifts a section of the Bishop's Road Canal Bridge, as the culmination of the salvage operation, March 2004.
[© Steven Brindle]

The following abbreviations have been used here and in the Bibliography:

BUL   Bristol University Library
GWR   Great Western Railway
NRNRG Network Rail National Records Group, York (formerly Network Rail Western Region Plans Room (NRWRPR), formerly GWR. I have used the original GWR reference numbers wherever possible, as the only consistent numbering system.
NA Rail National Archives (formerly Public Record Office) British Rail Archives.

Preface

1   The main elements of the proposed World Heritage Site would be Paddington Station, the Hanwell Viaduct, Maidenhead Bridge, Reading Station, the GWR's original village and workshops at Swindon, Twerton and Box Tunnels, the viaducts into Bath, Bath Station, and Bristol Temple Meads Station.

Chapter 1

1   Vaughan 1991, 56.
2   MacDermot 1964a, 1–2. MacDermot's official history, first published in 1927 and revised in 1964 by C R Clinker, has been followed by others, notably Nock 1962 and Booker 1977.
3   Booker 1977, 16; Stretton 1901. A railway from the Mersey to Birmingham was initially promoted by a group of Birmingham businessmen in 1824, but they were unable to raise the necessary finance and sought Liverpudlian support. Liverpool interests provided most of the money for the Grand Junction Railway, determined the choice of Joseph Locke as engineer, and dominated the board, Stretton 1901. The London & Birmingham Railway was founded by businessmen from London and Birmingham in 1830, who were swiftly joined by Liverpool interests, who came to dominate the company's 'Birmingham Committee', Bailey 2003, chapters 1 and 2.
4   MacDermot 1964a, 2–3.
5   Ibid, 2–3; Vaughan 1991, 44–5.

6   MacDermot 1964a, 2.
7   Vaughan 1991, 45.
8   The question of which was the first true railway is something of a conundrum, and depends on what is meant by the word 'railway'. Horse-drawn tramways had been in use at collieries for some decades, and in the early 1800s the first horse-drawn railways had been built independently of collieries: the Surrey Iron Railway (1803) and the Cromford & High Peak line. The Stockton & Darlington was the first railway designed for steam-hauled traffic but was primarily intended for goods. The Liverpool & Manchester was the first railway designed to operate regular passenger services.
9   Buchanan 2002, chapters 1 and 2; Brindle 2005, chapter 1.
10   Vaughan 1991, 1–41; Buchanan 2002, 22–7, 43–62; Brindle 2005, chapter 2.
11   Vaughan 1991, 47; Rolt 1957, 51–68; Brindle 2005, chapter 4.
12   Brunel 1870, 65.
13   Brindle 2005, chapter 4.
14   MacDermot 1964a, 4, says that the name was changed at the first joint meeting of the two committees. Vaughan (1991, 49) says that the meeting in question was on 22 Aug. Vaughan also draws attention to Marc Brunel's fondness for the word 'great', suggesting that Brunel may have been responsible for the new name.
15   Ibid, 49.
16   NA Rail 250/2, GWR board minutes 1835–41, 15 Sep 1835.
17   MacDermot 1964a, 5.
18   Ibid, 5–15.
19   The principal exceptions would appear to be the London & Southampton (1835) and the Eastern Counties Railway (1836), which were on a smaller scale. Reed (1996, 3 and 10–11), remarks that most of the money for the London & Birmingham Railway came from Liverpool, though its management passed to London, and that the Grand Junction Railway was both financed and run from Liverpool. A large part, perhaps as much as 40 per cent of the GWR's share capital came from Lancashire, too. The predominance of northern engineers and Lancashire money in the early years of the railway industry

is an important but under-researched subject.
20   Conder 1983, 118–23. F R Conder, a railway engineer, remarked in his memoirs on the contrast between Brunel's way of doing things and everyone else's: 'one was the scientific knowledge of a pupil of the "École Polytechnique", which contrasted rather sharply with the general and time-honoured English system of the "rule of thumb"'.
21   Buchanan 2002, 65.
22   MacDermot 1964a, 16–17. George Stephenson had been working at Killingworth Colliery, which had a tramway with a 4ft 8½in. (1.4m) gauge, not of his designing: this was where he built his first locomotive. The Stockton & Darlington Railway was based on this specific model.
23   Ibid, 26–8, 39.
24   NA Rail 1149/4, Brunel letter book, 1836, 262. Brunel wrote to the engineer Nicholas Wood with calculations of the additional cost of making the line wider, arriving at a total of £151,840.
25   MacDermot 1964a, 26–9. Much has been written on the subject of the broad gauge: *see* Bryan 2000, 37–52.
26   MacDermot 1964a, 17, quotes Brunel's testimony to the Gauge Commission of 1844–5: 'I think the impression grew upon me gradually, so that it is difficult to fix the time when I first thought a wide gauge desirable, but I daresay there were stages between of wishing that it could be so and determining to try and do it.'
27   Brunel's letter is quoted in full in MacDermot 1964a, 17–19.
28   NA Rail 250/2, 18, GWR board minutes for 29 Oct 1835.
29   NA Rail 250/2, 91, GWR board minutes for 27 Dec 1835; MacDermot 1964a, 35–47; Vaughan 1991, 108–19.
30   NA Rail 250/2, 9, GWR board minutes for 15 Sep 1835.
31   Buchanan 2000, 15–24.
32   Vaughan 1991, 62–5, 128–33; Rolt 1970, 105–7.
33   Many of Brunel's sketchbooks survive in the possession of Bristol University Library. Several sketches from this source are referred, *see* pp 27–8, 32–4, 38–40.

34 The GWR's archives were transferred to British Rail, the paper archive largely going to the National Archives at Kew, the plans remaining in the plan room at Paddington, until they were transferred to a Western Region plan room at Swindon. Network Rail has recently united its various regional plan rooms in a national record centre at York, referenced here as Network Rail National Records Group (NRNRG).

35 Conder 1983, 118, stresses Brunel's dictatorial style and control over every detail.

36 Vaughan 1991, 64–5, gives a number of examples.

37 Latimer 1887, quoted in Binding 2001, 11–12.

38 Conder 1983, 118–23. F R Conder was himself a railway engineer and contractor, and his detailed account of Brunel's dictatorial methods seems to have been written from first-hand testimony.

39 Ibid; Vaughan 1991, 68–70. Vaughan emphasises Brunel's tendency to use small and under-capitalised contractors, but this was not always so: he used the McIntoshes, Grissell & Peto, and Peto & Betts, amongst the best-organised builders of the age. Ranger was a case in point. He lost his claim against the GWR for compensation in 1843, died insolvent, and his creditors kept the case going in Chancery until 1859 (*see* Binding 2001, 23). Vaughan (1991, 134–9) gives a very interesting account of Brunel's arbitrary and scarcely legal treatment of the Scottish firm of Hugh and David McIntosh. Brunel withheld payments on a huge scale, the McIntoshes sued, and the case was only resolved in 1865, when the Lord Chancellor found for the McIntoshes, and ordered the GWR to pay them £100,000 with 20 years' accrued interest and costs. Vaughan gives a truer and more rounded picture than is supplied by Rolt (1970) or MacDermot (1964a, 55): the former never mentions the McIntosh case, the latter brushes it aside with a quip about 'Jarndyce & Jarndyce'.

40 NA Rail 250/64, 14. At the first meeting of the GWR proprietors in Oct 1835, it was argued that 'the moderate amount of excavations, not exceeding ten million cubic yards for the whole line' would tend to keep the costs down; it is not known how far this prediction was borne out.

41 NA Rail, 250/2, 43. It was named after Lord Wharncliffe, chairman of the committee, on the GWR's Act of Incorporation in the House of Lords. His coat of arms was carved on the centre of the south face.

42 NA Rail 250/82, Brunel's reports to the board, 71, 116–9, 141; MacDermot 1964a, 48–9.

43 MacDermot, 1964a, 30–3.

44 Ibid, 48–4.

45 Binding 2001, 71.

46 NA Rail 250/82, 121–5; MacDermot (1964a, 63–71) says that a hundred men lost their lives in the Box Tunnel, though it has been suggested (Rolt, 138–9) that this was the loss of life involved in the construction of the whole line.

47 MacDermot 1964a, 72.

48 Buchanan 2002, 76; Booker 1977, 39–40. There is some uncertainty about the Bristol and Exeter Railway board – John Binding (2001, 119) observes that, surprisingly, none of the original Bristol & Exeter directors were on the GWR board.

49 Simmons 1971, 6–8; Brindle 2005, chapter 6.

50 Buchanan 2002, 58–62; Brindle 2005, chapter 6.

51 Buchanan 2002, 52–62 gives a good account of the Bristol Dock Company's slothfulness and want of enterprise: both the *Great Western* and the *Great Britain* were too large to use Bristol's Floating Harbour, and could only be got out of it after their construction by, in the one case, the paddle wheels being taken off, and in the other, the entrance lock being widened. Clearly, the harbour was too small to handle such ships on any regular basis and Brunel's response was to urge the Dock Company to build deep-water facilities at Avonmouth. Their refusal to take any such action drove the Great Western Steamship Company to operate the *Great Britain* out of Liverpool and discouraged them from commissioning further ships. When the *Great Britain* ran aground in 1846, the cost of refloating her finished off the Great Western Steamship Company. By this point, the future had effectively been handed to Liverpool and to Edward Cunard.

52 MacDermot 1964a, 100–76.

53 Buchanan 2002, 78–9.

54 MacDermot 1964a, 196–207; Booker 1977, 103–7.

55 MacDermot 1964a, chapter 15; NA Rail 1149/4, Brunel's letter book for Mar to Nov 1836, has a number of letters on the subject.

56 MacDermot 1964a, 372–97; Vaughan 1991, 103, 116.

57 MacDermot 1964a, 29, 397–401. The *North Star* was delivered in Nov 1837. A 'sister' engine, the *Morning Star*, also commissioned by the New Orleans Railway, was converted and delivered to the GWR in Dec 1838, by which time the railway had opened.

58 Ibid, 329.

59 Vaughan 1991, 119–25.

60 NA Rail 250/2, 97–8. The board authorised the commissioning of 20 new locomotives on 10 Jan 1839; MacDermot 1964a, 397–401.

61 Cattell and Falconer 1995.

62 MacDermot 1964a, 335; Booker 1977, 78.

Chapter 2

1 'Bristol and London Railway – At a numerous and respectable meeting of Inhabitants of Bristol and its Neighbourhood, desirous of assisting in the Establishment of a Railway to London held (pursuant to public advertisement) at the Guildhall, Bristol, on Tuesday the 30th day of Jul 1833 – Report of the Provisional Committee', copy in the Science Museum, London, quoted in Tutton 1999, 3, note 1. Mr Tutton's scholarly and thorough book has been invaluable in the preparation of the present work.

2 NA Rail 1014/1, album of prospectuses for the GWR.

3 NA Rail 252/2.

4 Tutton 1999, 5–6.

5 NA Rail 250/1, 45.

6 MacDermot 1964a, 9.

7 Ellaway 1994, 1–3.

8 The discussions are briefly recorded in the GWR's board minutes, NA Rail 250/2, 8, 16, 27; and in the report to the general meetings of the proprietors of 26 Feb 1836, NA Rail 250/64. More can be gleaned from an abstract of the minutes of the GWR London Committee for 1833–8, NA Rail 250/83: unfortunately, the minute books for these years are missing, lent to a firm of solicitors at some point and never returned. *See also* Tutton 1999, 7–8. The London & Birmingham's side of the negotiations can be read in the minutes of their London Committee for 1835, NA Rail 384/41, and 1835–6, NA Rail 384/42.

9 NA Rail, 250/82, I K Brunel's reports to the board of directors, 19 Nov 1835; Tutton 1999, 13–14. The London & Birmingham Railway operated the stationary steam engine system until 1845, when they began to run locomotives into Euston in the usual way.

10 NA Rail 250/64, reports of general meetings of the proprietors, for 26 Feb 1836. The directors explained that, their request to purchase or to acquire a 21-year lease having been refused, 'it had come to their notice that a large and influential body of distant proprietors were urging the Directors of the London & Birmingham Railway to retain in their hands the power of breaking off any arrangement that might be made with the Great

Western Railway Company at the shortest possible notice, as well as from a hope of afterwards obtaining higher pecuniary terms, as from a fear that there would not be room for the traffic of both the lines and of their respective branches at the depot'.

11   The letter is quoted in full in MacDermot 1964a, 17–19.

12   NA Rail 250/2, GWR board minutes, 29 Oct 1835.

13   NA Rail 250/83, abstract of minutes of the GWR London Committee, 1833–8, meetings of 21 Nov 1835, 24 Nov 1835 and 26 Nov 1835.

14   NA Rail 250/82, 49, Brunel's report to the board of 9 Feb 1836. This stage of events is covered in detail in Simmons 1971, 18–23. Gibbs was one of the GWR directors, Tutton 1999, 16–18.

15   NA Rail 250/64.

16   NA Rail 250/82, 16–18. Hardwick's design seems to have represented exactly what he had agreed with Brunel on 19 Nov 1835.

17   Sir Nikolaus Pevsner (1976, 227), for example, completely misunderstood Hardwick's design: 'the functionalist is bound to deny that Hardwick worked on the purest principle of functionalism. This display of solemn monumentality is a screen: it is no more. That is: the rails and the necessary buildings alongside of them were not even on axis with the propylaea.'

18   NA Rail 384/43, 75ff, minutes of the London Committee of the London & Birmingham Railway, 14 Sep 1836. Stephenson advised them that 'more accommodation could be obtained for passenger traffic by a double station than by any possible arrangement of a single station'.

19   NA Rail 384/43, 110, 12 Oct 1836: Hardwick laid detailed plans before the committee, which were approved, and expressed his readiness to go to Liverpool to explain the plans to the 'Lancashire Directors'. NA Rail 384/43, 175, 7 Dec 1836: Hardwick reported that he had shown the designs to 'non-resident Directors and some of the principal proprietors' at Liverpool on 5 Dec 1836, and had received their 'unqualified approbation'.

20   NA Rail 384/155, contract for the portico, lodge and gates of Euston Station, £39,850. NA Rail 384/156, contract for arrival and departure stages, iron-slated roofs, drains, fence walls, paving roads and laying permanent way at Euston, £13,890, both contracts with William and Lewis Cubitt.

21   MacDermot 1964a, 32–47.

22   NA Rail 384/43, 75ff. The minutes for the London & Birmingham Railway's London Committee of 14 Sep 1836, which decided to go ahead with the 'double station', do not mention the GWR at all, but do make it clear that the London & Birmingham thought that half of the planned station would easily cope with their existing traffic, with room for growth by lengthening the platforms. So the idea that Euston was built with a view to the London & Birmingham taking control of the GWR is suppositious, but not unreasonable in the light of what happened in 1836–8.

23   MacDermot 1964a, 22.

24   Richardson 1999 (unpub report).

25   G Gutch, plan of the parish of Paddington, 1828, Westminster Archives.

26   Tutton 1999, 19–20; Simmons 1971, 20.

27   NRNRG GWR 7494, includes plans of the site, including the property plan 'attached to the agreement with Mr Cockerell', undated. Tutton 1999, Appendix I, 51–3, lists all of the landowners and lessees from whom the GWR bought property, quoted from 'Great Western Railway Book of Reference' (attached to their Parliamentary Deposited Plan, Nov 1836), Metropolitan Record Centre, London.

28   NA Rail 1008/64. Land at Paddington – notes of a committee appointed by the Paddington Parish Vestry on the application of the GWR Directors.

29   Paddington Parish Vestry minutes, 4 Feb 1837, Westminster Archives, quoted in Tutton 1999, 22, note 36: 'Brunel and Saunders attended the meeting of Paddington Vestry on 4 Feb 1837, the main points of the agreement being (1) that the GWR would built a 60 foot [18m] wide road – the present Bishop's Bridge Road, including the bridge over the railway and canal; (2) that the railway would be below street level; (3) the company would pay compensation for wear and tear to the roads about; (4) the GWR would purchase some parish land, the 'Bread and Cheese Lands', for £1,200; (5) the GWR would built all sewers caused by changes in level of the roads; (6) the new bridge over the railway at the intersection of Black Lion Lane (Lord Hill's Bridge) should not be less than 40 feet [12m] wide; (7) the new Bishop's Bridge Road and bridge should be made within 18 months of the passage of the Paddington Extension Act.'

30   NA Rail 250/83, 10, minutes of GWR London Committee.

31   Tutton 1999, 26.

32   Ibid, Appendix III, 60, estimated costs of the extension to Paddington, from a summary read at the half-year shareholders' meeting on 15 Aug 1838.

33   Tutton 1999, 22, quoting Paddington Parish Vestry minutes for 1836, Westminster Archives.

34   BUL, Brunel Collection, sketchbook labelled 'GWR 1836', 47–9; NRNRG GWR 7494. Both the latter plans are unsigned and undated, but they relate fairly directly to Brunel's sketchbook.

35   NA Rail 1005/82, extract from minutes of the GWR General Committee, 6 Nov 1845.

36   NA Rail 250/64, half-yearly meeting, 31 Aug 1837.

37   Tutton 1999, Appendix II, 54, quoting from NA Rail 272/38–42, London General Ledgers, and NA Rail 272/67–71, London General Journals, 1836–45.

38   NA Rail 1149/44, GWR Tenders, 35, tenders for 'London Terminus – various works': this gives the schedules of prices for the contracts, but not the main contract sums: 'retaining walls on Lord Hill's ground' (Sherwood, 23 May 1837); Black Lion Road Bridge (Sherwood, 19 Jun 1837); Westbourne Bridge (Grissell & Peto, 26 Jun 1837); Ranelagh Bridge (Grissell & Peto, 2 Oct 1837); Paddington Sewer (Grissell & Peto, 4 Mar 1837); Arrival and Departure Sides (Grissell & Peto, 17 Nov 1837); Paddington Canal Bridge (Sherwood, 27 Oct 1838). NA Rail 1149/8, one of Brunel's volumes of 'Facts', has Grissell & Peto's schedule of prices for the terminus buildings, dated 17 Nov 1837, listing the unit cost of all the materials: 24s per cubic yard of brick set in mortar, as against 31s per cubic yard set in cement, for example. NA Rail 250/82, board minutes for 6 Jul 1837: Brunel described both firms as 'highly respectable brick layers and excellent workmen'.

39   Tutton 1999, Appendix II, 54.

40   MacDermot 1964a, 29–30.

41   Tutton 1999, Appendix III, 60, quoting NA Rail 250/64, 6th half-yearly general meeting, 15 Aug 1838.

42   Tutton 1999, Appendix II, 56–7.

43   BUL Brunel Collection, 'Miscellaneous Sketchbook' (undated), ff. 28–9; NRNRG GWR 14891–8, has an unsigned, undated drawing for the gates and lodge, marked 'Great Western Railway – London Terminus'.

44   NRNRG, GWR 14909–21, is a fine series of coloured drawings for the Paddington goods shed, showing a triple-roofed timber-framed shed. These look similar, but not identical, to contemporary views of the Paddington goods depot, but they are unsigned and undated, and it is possible that they relate to a reconstruction of c 1845–6. NRNRG, GWR 14891 has two unnumbered and unlabelled drawings, which also appear to be for the first goods shed.

45 E C Matthews (1916b, 205) gives the dimensions of the departure platforms as, no. 1, 240ft (72m) long by 17ft 6in. (5.3m) wide, and no. 2, 235ft (71.6m) long by 17ft (5.2m) wide.

46 Ibid, gives the dimensions of the arrival platforms: no. 1, 255ft (77.7m) long by 25ft (7.6m) wide, no. 2, 340ft (103.6m) long by 34ft (10.4m) wide.

47 Tutton 1999, 55; NRNRG, GWR 7494, includes an undated plan for the building.

48 NA Rail 1149/5, Brunel letter book, 150, 233, letters to Philip Hardwick, architect of the Reading engine house. The first letter, dated 28 May 1839, says that the traverser in the engine house would not have to take a carriage but the one outside would. The second, dated 3 Oct 1839, directs Hardwick to make the foundations for the engine house traverser 25ft (7.6m) long. Binding 2001, 109–10, gives a good account of the Bristol traversers, quoting a description of them by one A J Dodson in the *Proceedings of the Institution of Civil Engineers*, **III**, 1844. It is not clear whether the traversers formed part of the original design, or when exactly they were installed. A key feature of the design seems to have been that a part of the framework rose, actually lifting the carriage or wagon off the rails to keep it immobile while the traverser was shifted sideways.

49 One of the traversers at Reading, referred to earlier, was in the engine house: one might think that locomotives would be too heavy for this to be operated manually, but it is not clear how else it could have been worked at this date, prior to the development of hydraulics. Matthews (1916b, 203) thought that the Paddington traverser must have all been manually operated, as the station does not seem to have had a steam engine at this date.

50 Wilson 1972, 27–8.

51 Thompson's engine house at Derby was demolished in 1985, with what remained of his trijunct station: Pevsner 1978, 176. Gooch's first sketch design for the GWR's locomotive department at Swindon, which accompanied a letter to Brunel in Sep 1840, also included a round engine house, but this was never built: Cattell and Falconer 1995, 6–10.

52 NA Rail 1149/9, 201, sketch by T A Bertram dated 1 Sep 1840.

53 MacDermot 1964, 78.

54 Westminster Archives, detached copy of steel engraving, dated 1843 and probably from *The Illustrated London News*.

55 NA Rail 1008/64. Letter from Paddington Parish Vestry to Charles Saunders, 13 Mar 1841.

56 MacDermot 1964a, 52. First-class passengers had a choice between sitting in the 'posting carriage', with upholstered seats, and riding in their own carriage on a carriage truck, while second-class passengers had to choose between a coach with wooden seats and a roof but no glass in the windows, and an unroofed open truck.

57 MacDermot 1964a, 333.

58 Ibid, 331.

59 Ellis 1975, 4–9.

60 Tutton 1999, 43.

61 Ibid, 43–6; MacDermot 1964a, 325–6.

62 Quoted in MacDermot 1964a, 326.

63 Ibid, 326–8.

64 NA Rail 1005/82, typed transcript, extracts from the minutes of the General Committee, 6 Nov 1845.

65 NRNRG GWR 14909–21 is a set of fine large coloured drawings for a timber-framed shed with wrought-iron trusses. They are unsigned and undated, but are labelled 'Paddington Goods Shed'. These may be the unexecuted proposal of 1845–6.

66 NRNRG GWR 7494.

67 Brindle and Tucker 2004.

Chapter 3

1 NRNRG, GWR 14891–2, drawings for new one-storey carriage shops adjoining the existing paint shop, signed Joseph Griffiths, undated, but dated to 1847 in the catalogue; 14893–8, undated drawings of new timber-clad carriage sheds and engine house, with variant brick and timber clad elevations, seem likely to be of the 1840s, but this is not certain.

2 The original volume of half-yearly reports covering 1850 appears to be missing. I have used typed transcripts in NA Rail 1005/82, miscellaneous historical documents.

3 Vaughan 1997, chapters 12 and 13; Binding 2001, 51.

4 Jackson 1969, 39–42; Hobhouse 1976, 41.

5 NA Rail 1005/82, transcripts of GWR half-yearly reports.

6 Meeks 1957, 39.

7 Binding 2001, 53–4.

8 Meeks 1957, 61–2.

9 BUL, Brunel Collection, Small Sketchbook 22, fol 7.

10 BUL, Brunel Collection, Large Sketchbook 3, fols 1, 7.

11 Idem. The dimensions do not quite tally between the two sketches. For example, on fol 1 the central concourse is clearly marked as 450ft (137.2m) long, on fol 7, adding up a number of measurements it seems to come to 420ft (128m): this

is perhaps only to be expected in rough initial sketches like this.

12 BUL, Brunel Collection, Large Sketchbook 3, fol 5, 10. These sketches relate closely to the tiny vignette on fol 1 of the same sketchbook.

13 Binding 2001, 78–102. Foyle 2004, 88–90. The Bath Station roof is something of a puzzle. John Binding (2001, 153), notes that no drawings for it have been found. However, there are sketches in one of Brunel's sketchbooks, BUL Brunel Collection, GWR Sketchbook 9, 19–20, probably of 1840, in which Brunel shows heavy tie-rods tying it across the mid-span. Oddly enough, these tie-rods do not appear in the view published by J C Bourne (Bourne 1846).

14 Ibid, 78–102. Mr Binding estimates that 80–90 per cent of the stiffness in the Bristol roof comes from the ironwork. Heavy horizontal timber crosspieces were also added over the aisles, apparently to help hold the roof down and counteract the outward forces: it is clear, though, that the walls and columns have deflected over time.

15 Colquhoun 2003.

16 Sutherland 1989, 107–26.

17 Addyman and Fawcett 1999, chapter 5; Meeks 1957, 59–60.

18 Meeks 1957; Turner 1850, 204.

19 Brunel was also a member of the Machinery Section Committee, and the chairman and reporter of the judges for Class 7 exhibits, on civil engineering, architecture and building technology. Rolt 1970, 230.

20 Colquhoun 2003, 164–6; McKean 1994, 4–9.

21 McKean, 1994, 10–13; Buchanan 2002, 188, quotes Brunel writing in May 1850 that he was 'brim full of the details of dome and specification'. Unfortunately, no detailed drawings of the dome are known, so we cannot assess Brunel's approach to this novel problem.

22 Cross-Rudkin and Chrimes (eds) 2008, entry on Paxton by James Sutherland, 608–11; Colquhoun 2003; McKean 1994.

23 Colquhoun, 2003, 164–72.

24 McKean, 1994, 15–20.

25 NA Rail 250/4, GWR board minutes, 248.

26 Hobhouse 1976, 40–2.

27 NA Rail 250/4, 259.

28 NA Rail 250/4, 274.

29 BUL, Brunel Collection, Large Sketchbook 3, fol 1.

30 BUL, Brunel Collection, office diaries for 1850, 1851.

31 NA Rail 250/4, 254, GWR board minutes. The board also discussed their need for additional capacity in view of the Great Exhibition, due to open on 1 May 1851.

This might have encouraged them to go ahead but can hardly have been a determining factor as the new station could not have been ready in time for the exhibition.

32 Ibid.

33 BUL, Brunel Collection, Large Sketchbook 3, fols 11–13.

34 BUL, Brunel Collection, office diary for 1850.

35 NA Rail 250/4, 298.

36 Binding 2001, 53–77.

37 Hobhouse 1976, 40–2.

38 Ibid.

39 Ibid, 310.

40 BUL, Brunel Collection, office diaries for 1850, 1851.

41 Vaughan 1991, 62, 68–9, 81–2, 119–20, and the office diaries.

42 The whereabouts of the original MS is unknown. There is a typescript in NA Rail 1005/82, and it is quoted in full in Rolt, 231–2.

43 BUL, Brunel Collection, office diary for 1851.

44 Ibid. Brunel saw Fox and Henderson at 10.15 am, Bertram at 11.45 am, Bertram again at 4.30 pm, as well as having meetings about the West Cornwall, South Wales, Oxford & Worcester and Vale of Neath railways, the Clifton bridge, the Chepstow bridge, Sunderland Docks, the Eastern Steam Navigation Company, the Galvanised Iron Company and the Vulcanized Rubber Company.

45 NA Rail 250/4, 328; NRNRG, GWR 13320, contract drawings for Fox Henderson's first contract.

46 *Oxford Dictionary of National Biography*, **7**, 533–4, entry on Sir Charles Fox by R Thorne; Cross-Rudkin and Chrimes (eds) 2008, entry on Charles Fox by D Gwilym M Roberts, 310–15, and entry on John Henderson by Tom Swailes, 397–8; Conder, 10–11.

47 Weiler 1987 (unpub report), 144. James Sutherland considers that it was the use of rolled wrought-iron angles and tees here, which was particularly significant: pers comm.

48 *Victoria County History Staffs* **17**, 1976, 110–11; Sutherland, 1989, 116.

49 Weiler 1987 (unpub report) 143–61; Sutherland 1989, 116.

50 Cattell and Falconer 1985, 32–4.

51 *Victoria County History Staffs* **17**, 1976, 111.

52 Sutherland 1975, 69–72. It used to be thought that the Oxford building was built from 'spare parts' for the Crystal Palace: James Sutherland demonstrates that it was similar (built to the same 24ft (7.3m) module), but not identical. In particular, the wedged fixings of the

Crystal Palace's beams and trusses were replaced by a more solid bolted fixing. The Oxford building has been dismantled and reconstructed at the Buckinghamshire Railway Centre, Quainton Road.

53 Weiler 1987 (unpub report), 143.

54 Cowper 1854.

55 NA Rail 1005/82, *The Josser*, 20 Mar 1889: this appears to have been an unofficial, privately circulated staff newsletter.

56 BUL, Brunel Collection, Large Sketchbook 1, fol 40, Large Sketchbook 3, fols 3, 15, 35.

57 NA Rail 1005/82, typescript copy of the specification for Fox, Henderson's second contract for Paddington passenger station, dated 8 Nov 1852. 'Generally the curved roof is to be similar in its construction and character to that already executed at Paddington New Station which is referred to as an example of the work required and in all cases unless otherwise specified the details of construction and description of materials and workmanship are to be similar to those adopted in like positions in the work already executed. Materials. Wrought Iron – ribs, girders, angle irons, etc. Cast Iron – columns, plinths and baseplates and ornamental castings. Paxton Glass roofing in part – patent rolled plate 26 oz [793gm]. Galvanized Corrugated Iron Roofing in part – no. 18 gauge. York Stone bases for columns. Timber – red pine or memel and deal battens for gallery. Arched spans – to be 102 feet 6 inches [31.2m] and 68 feet [20.2m] respectively. Height above rails – 54 feet 6 inches [16.6m] and 46 feet [14m] respectively. Estimated quantities – 300 cubic yards [223cu m] of brick work, 100 cubic yards [76.5cu m] of concrete, 50 tons [50.8 tonnes] of wrought iron girders, 20 tons [20.3 tonnes] of cast iron columns, brackets, etc. 8 tons [8.1 tonnes] of lead guttering, 4,582 cubic yards [3,503cu m] of excavation. Maintenance of Works – twelve months after works constructed under the contract have been executed.' There is something of a mystery about the figures of 50 tons for the wrought iron and 20 tons for the cast iron, which look far too small.

58 BUL, Brunel Collection, Large Sketchbook 3, fols 22, 28, 37. Fol 22 shows the transept with a semicircular arched profile, as executed.

59 The southern span was given tie-rods early in the 20th century, which it retains.

60 Binding 2001, 86, illustrates the 'tie-rod' design, observing that the decision to omit the rods 'had a significant effect on the subsequent behaviour of the roof in service, by introducing horizontal forces at the walls'.

61 Hitchcock, 1951 and 1972; Dixon and

Muthesius, 1978, 105; Pevsner 1950.

62 This assumes that Brunel used his sketchbooks sequentially: on fol 11, there is the first sketch plan for the revised design, as built, which seems to confirm that fols 5 and 10 relate to the rejected central concourse scheme.

63 *Journal of the Royal Institute of British Architects*, 3rd series, **11**, 19 Dec 1903, 116. I am grateful to Mr Robert Thorne for this information. Fowler signed some of the very fine drawings for railings and gates on Eastbourne Terrace, NRNRG, GWR 13315A.

64 NRNRG, GWR 10270.

65 Cross-Rudkin and Chrimes (eds) 2008, entry by Steven Brindle on Matthew Digby Wyatt, 861–2.

66 Ruskin 1849, 36–7.

67 Review of the *The Seven Lamps of Architecture* in *The Journal of Design*, **2**, 1849–50, 72. The review is anonymous but attributed to Wyatt by Sir Nicholas Pevsner on the grounds that the opinions are close to those expressed in other articles by him.

68 Wyatt 1850.

69 NRNRG, GWR 420.

70 NA Rail 250/5, 24.

71 Ibid, 37.

72 BUL, Brunel Collection, Large Sketchbook 4, 15–16, dated 29 Jan 1851.

73 NA Rail 250/5, 53. The drawings Brunel refers to here are probably those now in NRNRG 13324. These relate to the area north of the viaduct leading up to the high-level yard, about two-thirds of the site. The southern part of the goods depot was not designed until 1853: NRNRG 16122, 13304, 17842.

74 NA Rail 250/5, 73.

75 Ibid, 96.

76 Ibid, 211.

77 Ibid, 243.

78 Ibid, 251.

79 Cross-Rudkin and Chrimes (eds) 2008, entry on Armstrong by R W Rennison, 25–30; McKenzie 1993, chapters 3 and 4; BUL, Brunel Collection, office diary for 1851. Armstrong is misnamed as W J – he was W G – but there seems to be no doubt that it was indeed him. Brunel met him on 3 Feb and 14 Mar, then there was a hiatus, then meetings on 8, 10 and 11 Apr, 3 and 19 May, then one more meeting on 15 Jul. NRNRG, GWR collection has numerous drawings of hydraulic machinery for Paddington by Sir William Armstrong & Co, for example 13309, 13310, 13312. Further rolls of drawings for machinery are listed at 18674–8 and 18679–82, but appear to be missing. *See* chapter 5, for a fuller discussion of these documents.

80 NA Rail 250/5, 258.

81 Ibid, 377.

82 Ibid, 398.

83 NA Rail 1008/35, Brunel to Saunders, 17 Mar 1852.

84 NA Rail 250/5, 405.

85 Owen Jones was on the Executive Committee of the 1851 Great Exhibition, and was responsible for the much-acclaimed colour scheme for the Crystal Palace. This seems likely to be how Wyatt and Brunel knew him. There do not appear to be any references to Jones in the GWR records, presumably because the company was never responsible for employing him. References to his work at the station come from the *Athenaeum* magazine, in turn referring to William Powell Frith's famous painting *The Railway Station*. Carol Flores' biography of Owen Jones (2006) has no references to Jones' scheme for Paddington, and any drawings he made for it seem to be lost without trace.

86 NA Rail 1005/82, Saunders to Brunel, 15 Apr 1852.

87 Levien 1930, article about the painting. *See* NA Rail 1005/56, file relating to the painting.

88 M D Wyatt would have known Alfred Stevens, like Owen Jones, through his work on the Great Exhibition. It is thought that Stevens's designs for the royal waiting room probably date from 1854 (Beattie 1975, 32–3). Three letters on the subject from Stevens to Wyatt are preserved, Victoria and Albert Museum Library, E996–1914. Stevens produced a number of variations of his design, quoting a price of £15 for each of the eight sides of the ceiling. He asked for payment in 1856, noting that 'it seems unlikely that the scheme for decorating the Queen's waiting room at Paddington Station will ever be carried out'. Wyatt replied on 29 Mar 1856 enclosing a cheque for £10 and both drawings, regretting that they 'had not been able to succeed in getting the decoration carried into effect'.

89 NA Rail 250/6, pp 6, 24, 31, 196 and 274–5. The Great Western Royal Hotel Company, thus formed, took a 42-year lease and ran the hotel until 1986, MacDermot 1964a, 269.

90 NA Rail 1005/82, directors' report to shareholders, 17 Feb 1853.

91 Ibid.

92 Brunel 1870, 86.

93 NA Rail 250/6, 41–2.

94 Ibid, 49, 71. A copy of the specification for Fox, Henderson's second contract, dated 8 Nov 1852, is in NA Rail 1005/82. The drawings for the second contract are in NRNRG, GWR 13320.

95 NA Rail 250/6, 71, GWR board minutes for 22 Jul 1852.

96 NA Rail 250/6, 229. Two of Gainsford's reports to the board are extant, for Jul and Aug 1854, NA Rail 1014/33.

97 NA Rail 250/6, 354.

98 Ibid, 358.

99 Ibid, 381–2.

100 Ibid, 414.

101 NA Rail 1005/82, Saunders to Brunel, 4 Apr 1853.

102 Vaughan 1991, 153–6, 215–7; Brindle 2005, chapter 6.

103 NA Rail 1005/82. 'During the last fortnight, several of the Board having themselves visited and inspected the buildings and offices, in which scarcely any human being is at work, they have taken it upon themselves, in the absence of Mr Brunel from London, to give the orders they deem requisite for insuring early use and occupation of the building. They will be glad that your clerks shall vacate the rooms within two or three days, which they have been using in the upper floor, in order that they may be made fit for the clerks of the Company … I am also to say that the inaction in all that relates to the new roofs for the Carriage Sheds oblige the Board very shortly unless they perceive a manifest change, to take that work into <u>their own hands</u> to be finished – for there is not the least appearance of its being completed – according to the present state of it.'

104 NA Rail 1008/35, Brunel's reports to the GWR board.

105 NA Rail 250/7, GWR board minutes for 1853–4, 45, 11 Aug 1853, record a decision that the arrival platform be made wider, and the intermediate platform reduced by the same amount. Ibid, 141, 27 Aug 1853, note that the 'Rooms for the reception of Her Majesty' were incomplete: the sub-committee (it is not clear which) were to obtain designs for furnishing it from Messrs Holland.

106 NA Rail 250/7, pp 45, 151, 155, 189, 201, 233, 247.

107 *The Builder*, **12**, 28 Jan 1854, 41.

108 *The Builder*, **12**, 3 Jun 1854, 290–1.

109 *The Builder*, **12**, 17 Jun 1854, 322–3.

110 NA Rail 250/7, 373, 387.

111 The directors were obliged to seek support for any major capital project from the GWR's half-yearly shareholders' meetings, at which estimates were presented. Thus it is possible to plot the rising cost of the station at six-monthly intervals in the reports to shareholders. The station cost £250,037 to the end of 1853, £318,089 by 30 Jun 1854, £411,828 to the end of 1854, £480,661 to the end of 1855, £563,875 by the end of 1856, rising to £608,825 by the end of 1857, as work continued on fitting out the goods yard etc. This was in addition to the money spent on the hotel, which amounted to £49,134 by 30 Jun 1854 and £59,965 by 30 Jun 1855.

112 Birmingham New Street was damaged by bombing in the Second World War and partly dismantled. The entire station was destroyed and replaced with the present concrete bunker in the 1960s.

113 Buchanan 2002, 139–44, letter quoted on 141.

114 *Victoria County History Staffs* **17**, 1976, 111; Evans, 1859, Appendix, XCIII–XCIV. I am very grateful to Mr Robert Thorne for this information.

115 Cross-Rudkin and Chrimes (eds) 2008, entry on Sir Charles Fox by Gwilym Roberts, 310–15; entry on John Henderson by Tom Swailes, 397–8; Foyle 2004, 88–90.

116 Cross-Rudkin and Chrimes (eds) 2008, entry on Francis Fox by Brian George, 315–16; entry on Wyatt by Steven Brindle, 861–2.

117 Vaughan 1991, 215–17; Binding 2000, 86–99.

118 Brunel 1870, 461–73. I am very grateful to Andrew Saint for this information.

119 Brindle 2005, chapter 6; Buchanan 2002, chapter 8; Vaughan 1991, 230–5, 245–69; Griffiths 2000, 112–27.

120 Wilson 1972, 76.

Chapter 4

1 MacDermot 1964a, 163–4 and 207–9. MacDermot used the portrait as the frontispiece illustration to Vol I of his *History*. Later, the painting was loaned to Reading Town Hall, and was unfortunately cut down in size. It has been acquired by STEAM, Museum of the Great Western Railway, Swindon.

2 Bills and Knight 2006, chapters 3 and 5; Wood 2006, chapter 7; Chapel 1982, no. 23, 87–92.

3 Wood 2006, chapter 7. Despite the complexity of detail in the painting, Frith only paid Stone £300 each for the copies. One of the Stone copies is now in a private collection, and the other is in the Royal Collection. A third copy is in Leicester Museum & Art Gallery: it does not seem to be clear whether this, too, is by Stone.

4 MacDermot 1964a, 223–6.

5 In 1854, interest payments represented about 32 per cent of all the GWR's spending (£252,822 out of £783,769), as against *c* 24 per cent for the London & North Western Railway (£398,781 out of £1,687,712) and 18.3 per cent for the Midland (£158,051 out of £862,573). I have not found figures for the 1860s, but

it seems unlikely that the GWR's position would have improved.

6   MacDermot 1964b, 19–22; Booker 1977, 84–6. The company weathered the storm by drastic economies, by raiding its (mercifully strong) cash flow, and by offering the shareholders preference shares instead of dividends.

7   Thorne 1978, chapter 4, especially 44–52.

8   *The Times*, 1 Sep 1935, letter from Mr Jasper Lambert, written in reference to a recent issue celebrating the centenary of the GWR. Mr Lambert recalled that he had known a Mr Woolaston, once a pupil of Brunel's. Woolaston once asked Brunel, 'I wish you would tell me, Sir, why you constructed Paddington Station at the level at which it stands today?' 'Well', replied Brunel, 'I will tell you. For one thing, by so doing, I was able to get a quantity of very good gravel, but I had another reason. I felt convinced that one day there would be an underground railway running around London at about that level.'

9   BUL, Brunel Collection, office diaries, 2 Nov 1852. Brunel was engaged with 'Messrs Barratt, Davis and another gentleman'.

10 MacDermot 1964a, 229–31.

11 Barker and Robbins 1975, 99–125. A long account of the opening of the Metropolitan Railway is given in *The Builder*, **21**, 10 Jan 1863, 21–3. *See also* Menear 1984, 7–8; Trench and Hillman 1984, 130–8.

12 Barker and Robbins 1975, chapter 4.

13 Ibid, 126–7.

14 Levien 1936; Barker and Robbins 1975, 127, 145–6.

15 Barker and Robbins 1975; Menear, 1984, 9.

16 Anon 1933d, 419–26, esp. 420.

17 NA Rail 267/6, general manager's report on suburban passenger traffic, 4 Feb 1869.

18 I am grateful to Mr John Lewis for information on these points.

19 NA Rail 267/230, report on a proposed railway between High Wycombe and Paddington, 8 Oct 1896; White 1963, 121. The new route ran from Old Oak Common to High Wycombe, where it met the existing line from Maidenhead, via High Wycombe to Oxford. The line thus provided a new direct route to Birmingham, via High Wycombe and Oxford, and was promoted as such. I am grateful to Mr John Lewis and Dr Gavin Stamp for information and advice on this point.

20 NA Rail 250/1661, special report on the working of traffic at Paddington by W Hart, 1 Aug 1881.

21 NRNRG, GWR 7494, new departure platform under the central span, Jul 1873; 24581, new departure platform, 1883; 27903, additional platform, 1889.

22 NRNRG, GWR 24562, dated 4 Jan 1878 shows the proposed new works; 24575 is a drawing for a new canopy over the milk platform, signed J Olander, 1882.

23 NRNRG, GWR Drawings Catalogue, II, lists plan 14949 as being for this.

24 NRNRG, GWR Drawings Catalogue, I, no. 421. The drawing itself seems not to be extant.

25 NRNRG, GWR 13316, unexecuted designs for a 20-bay range initialled ETA, 22 Apr 1872.

26 NA Rail 258/343, contains detailed specifications for the building dated 1873; NRNRG, GWR 17021, has contract drawings dated 1876.

27 NRNRG, GWR 17162, additional storey of offices, contract drawings signed by Kirk & Randall and dated 31 May 1879; 17840, 'additional offices', drawings for the eastwards extension, 1879–81; 17841, designs of 1880–1 for additional offices including part of the new top storey; contract drawings signed by Kirk & Randall, 23 Nov 1880.

28 NA Rail 258/343.

29 *The Engineer*, **52**, 16 Nov 1881.

30 The date at which the tracks were curtailed is not known. The 1867 Ordnance Survey map shows them going right up to the rear of the hotel; by the 3rd edition (1912), the track heads had been pulled back into the area of Brunel's main shed roofs.

31 NA Rail 258/343, various documents. NRNRG, GWR ms catalogue, lists drawings 16149–51 as being for these alterations (not seen).

32 NA Rail 280/31, 383, GWR board minutes for 10 Nov 1880; 450, minutes for 5 Jan 1881.

33 Saint, Pettit and Irlam 1992, 33. Thomas Edison also invented the incandescent light bulb, quite independently of Swan but several months later, in the United States.

34 *Engineering*, 21 May 1886, copy in NA Rail 258/508; NA Rail 250/165, minutes of the GWR Electric Lighting Committee. These rather haphazard minutes mostly deal with the legal and contractual difficulties the company encountered over the installation of this system, 1884–6; Tweedie 1906, 242 gives a rather sanitised account.

35 *The Electrician*, May 1886. The rival for this title would seem to be the system installed at the Grosvenor Gallery on New Bond Street in 1883.

36 MacDermot 1964b, 231. The Metropolitan Railway obtained an Act to run the Hammersmith & City Line by electricity in 1902, and in 1906 the GWR completed a power station at Park Royal to supply electricity for this, and also to supply Paddington, replacing the equipment at Westbourne Bridge.

37 Ellis 1959, 40.

38 White, 1987, 119.

39 MacDermot 1964b, chapters 9 and 10. I am grateful to Mr Peter Treloar and Mr John Lewis for advice and information about the end of the broad gauge. In some areas where there was mixed-gauge track the third rail was left in place to be taken up later.

40 Booker 1977, 10 and 103.

41 MacDermot 1964b, chapter 11.

42 Booker 1977, 175–82. I am very grateful to Mr John Lewis for information about Churchward and his achievements.

43 Booker 1977, 101–15.

44 NA Rail 250/343, GWR Traffic Committee minutes, 1903–5, 7. This was done at an estimated cost of £19,000.

45 Ibid, 149; GWR Traffic Committee minutes, 8 Jun. The general manager referred to the extended up relief line, but said that even so, 'additional platform accommodation should be provided forthwith. He submitted a plan which, while departing from the approved scheme of alterations, may be regarded as an improvement on it. The Proposal is, by utilising the space between no. 9 platform and the High Level Coal Yard, to provide two additional Main Line platforms with roadway between, and a short platform for suburban traffic. The plan also includes substituted provision for Cabs in South Wharf Road and the maintenance of the access between that Road and London Street, which will be the subject of an agreement with the Paddington Borough Council.' The estimate, of £93,000 for engineering works and £3,950 for signalling and telegraph works, was approved by the board on 9 Jun 1904, NA Rail 250/47, GWR board minutes 1903–5, 219.

46 NA Rail 250/344, Traffic Committee minutes 1905–6, 257, 8 Aug 1906. Referring to GWR board minute 12 of 8 Jun 1904, the general manager said that 'in working out Proposals for Paddington, it had been found desirable to make variations, and submitted an amended plan. In this scheme the accommodations on the platform level has been slightly rearranged and an access given from the end of the South Wharf Road, which will greatly facilitate the working of the Milk Traffic. The scheme also involves the

removal of the buildings on the Down Platform at Bishop's Road Station to enable an improvement to be effected in the line serving No. 12 platform and platforms Nos 10 and 11 extended. A new Waiting Room and Urinals will be provided at the West End of that Station, and a new footbridge and booking offices in lieu of the existing structure, while provision is also made for the erection of a length of covering in the Cab Yard.' The estimate had risen to £111,000 for engineering and £4,500 for signalling works. This was approved by the board the next day, 9 Aug 1906: NA Rail 250/48, board minutes 1905–8, 184.

47 MacDermot 1964b, 230; White 1906.

48 NA Rail 250/344, GWR Traffic Committee minutes 1905/6, 225, 13 Jun 1906, report to the committee to the effect that the original vote of £19,720 had lapsed due to the protracted negotiations with Paddington Borough Council, and a new vote of £24,000 was agreed to. NA Rail 250/48, GWR board minutes, 147, the board's agreement on 14 Jun 1906 to the new vote; 173, the board agrees on 19 Jul 1906 to the main contract for the reconstruction of the Bishop's Road Bridge, for £17,843, let to Messrs Jackaman. NA Rail 250/49, the contract for the reconstruction of Westbourne Bridge was let to Messrs A Handyside on 6 Aug 1908 for £37,562.

49 *The Railway Engineer*, Mar and Apr 1906 has articles on the completion of the building: I am grateful to Mr Peter Rance of the Great Western Trust for this reference.

50 Anon 1907a and b.

51 Matthews 1917a, 28–31.

52 NA Rail 250/49, GWR board minutes for 30 Apr 1909. The minutes are very terse, and do not make any explicit reference to reinforced concrete. No detailed discussion of the design has been found in the GWR archives in the NA.

53 NRNRG, GWR 50537, 21 drawings for Hennebique ferroconcrete construction by L G Mouchel & Co.

54 Cusack 1986, 183–95; Cusack 1987, 61–7.

55 Ibid.

56 Anon 1907b makes it clear that the GWR was using the Hennebique system.

57 Unpublished reports by Andrew Saint, English Heritage, Historic Analysis and Research Team, Historical research files, CITY 228, 28–31.

58 Matthews 1917a, 28–31; 1917b, 69–70.

59 NA Rail 250/49, 230.

60 NA Rail 250/50, 35. GWR board minutes for 16 Dec 1910, approved letting a contract for £6,750 to Messrs Jackaman, for excavation and removal of 45,000cu yd

(34,407cu m) of material beneath the superstructure for carrying the new goods and cab approaches.

61 NA Rail 250/50, 246. GWR board minutes for 7 Feb 1912, approved letting a contract for £18,285 to Messrs Holliday & Greenwood for the diversion of London Street and construction of new underground conveniences.

62 NA Rail 250/50, 282. GWR board minutes for 26 Apr 1912, approved letting a contract for £720 to Messrs Jackaman for building a retaining wall instead of the ordinary platform wall included in the contract. Contract no. 4.

63 NA Rail 250/50, 434. GWR board minutes for 10 Jan 1913, approved letting a contract for £820 to Messrs Jackaman, for cellarage accommodation and the cab ramp, including extra foundations and the facing of the cellars with glazed brick.

64 NA Rail 250/50, 488. GWR board minutes for 11 Apr 1913, contract for £945.

65 NA Rail 250/51, 127. GWR board minutes for 21 Nov 1913, approved letting a contract for £28,969 to Messrs Holliday & Greenwood, for construction of a roof over platforms 9, 10 and 11 at Paddington.

66 NA Rail 250/51, 145. Board minutes for 16 Jan 1914, approved letting a contract for £505 to Messrs Jackaman, for deepening the foundations of two concrete bases to main roof columns. NRNRG 52134, 2 drawings, contract 17, details of new steel columns to roof, Armstrong.

67 NRNRG, GWR 51526, contract 12, 12 drawings for roof over platforms 9, 10, 11, W Armstrong, Dec 1913; 52131, contract 14, 20 drawings for roof extension, Gleadow & Armstrong; 52141, contract 16, 4 drawings of details of end screen, Gleadow.

68 NRNRG, GWR 52149, 3 drawings of tie-rods to the new span, Gleadow. These are dated 1913, well before the engineers' report, so tie-rods would seem to have been envisaged but not installed.

69 NA Rail 267/304, report on Paddington Station roof, 9 Jul 1915.

70 NRNRG, GWR 52657, six drawings for new steel columns between platforms 2 and 3, Aug 1923; 52658, 5 drawings for new steel columns between platforms 7 and 8, May 1923, both sets of drawings by the Cleveland Bridge & Engineering Company.

71 Booker 1977, 118–19.

72 NA Rail 447.

73 They were: the Great Western Railway, the London, Midland & Scottish Railway, the London & North-Eastern Railway, and the Southern Railway.

74 MacDermot 1964b, 242–2. The new group was composed of the GWR with 3,005 miles (4,836 km) of line, 2 small English

companies, and 13 small Welsh lines ranging in size from the Cardiff Railway with less than 12 miles (19.3km) to the Cambrian Railway with 295 miles (474.5 km). *See also* Nock 1967, 1–2.

75 Booker 1977, 120–8.

76 Waters 1993 107–9.

77 Pole 1926, 23.

78 Booker 1977, 182–6.

79 Simmons 1968, 43.

80 NRNRG, GWR collection, bound typescript volume, T S Todd, 'The Future of Paddington Station', 1926.

81 A table of traffic numbers and revenue on the GWR shows total receipts at Paddington as £1,089,914 in 1903, £1,261,611 in 1913, £2,115,552 in 1923, peaking at £2,303,054 in 1928, and falling in each successive year to £1,886,281 in 1933. The figures for Paddington goods show a similar profile.

82 Nock 1967, 87–8, 95–6. This was part of a £4,500,000 programme of capital works affecting much of the GWR network, assisted by the Government under the 1929 Act.

83 Anon 1933c, 345–6; Culverhouse 1934; *The Daily Telegraph*, 11 May 1933; Shackle 1933.

84 Binding 2001, 149; Foyle 2004, 90. The work at Bristol was also funded under the 1929 Loans and Guarantee Act.

85 NRNRG, GWR 63400, 33 drawings for new offices on the departure side, May 1931, Culverhouse; former NRWRPR microfilm 13924–13938, formerly GWR 63588, steelwork drawings for new offices and hotel extension on the departure side, 1930, E C & J Keay.

86 NRNRG, formerly NRWRPR microfilm 14984, formerly GWR 64611, 14 architectural drawings and 127 steelwork drawings for a new extension of the departure side office block, undated, Culverhouse.

87 NRNRG, GWR 63653, three drawings for demolition of buildings for the arrival side roadway, Feb 1932, Culverhouse.

88 NRNRG, microfilm 1344–61, formerly GWR 63687, 17 drawings for new offices on the arrival side and over the Lawn, Sep 1932, Culverhouse; microfilm 4402–37, 28496–98, formerly GWR 64029, 32 steelwork drawings for new arrival side offices and the hotel extension, Oct 1933, J S Nicholas and E C & J Keay.

89 NRNRG, microfilm 33382–98, 33401–11, formerly GWR 63474, 27 drawings for additions and alterations to the GWR hotel, 1935, Culverhouse, E C & J Keay.

90 NRNRG, GWR 63818, 9 drawings for the reconstruction of roofing over the circulation area, Jan 1933, J S Nicholas.

91 NRNRG, GWR 63474; 64951, 19 steelwork drawings for alterations and additions in the old part of the hotel, undated, R G Sargent and A D Dawney & Co; Culverhouse 1934.

92 Anon 1935: information from Mr John Lewis.

93 NRNRG, GWR 63266, 6 drawings for contracts 5 and 6, new spur tunnel and retaining walls at the Bishop's Road Station, Jun 1931, Carpmael; 63708, 14 drawings for the partial rebuilding of the Bishop's Road Bridge, Sep 1932, Carpmael; 63769, 8 drawings for the Bishop's Road Bridge, 1932, Carpmael, Cleveland Bridge & Engineering Co.

94 NRNRG, GWR 63406, demolition of the Bishop's Road Station, undated, Carpmael; 63751, reconstruction of Bishop's Road Station, 1932, Carpmael; 63791, 26 drawings for excavation, demolition and reconstruction of west end of tunnel, Feb 1932, Carpmael.

95 Nock 1967, 97.

96 Anon 1933c, 352.

97 Owings 1973. I am very grateful to Andrew Saint for this information.

98 The GWR set up its first bus services in 1903, but sold 50 per cent of the company in 1928: Nock 1967, 82. I am grateful to Mr Peter Rance for pointing this out.

99 Simmons 1968, 40–1.

100 Wodehouse 1954, 84–5.

101 Nock 1967, 216, gives figures of 444 GWR employees killed, 155 reported missing, and 271 prisoners of war, to Mar 1944. No figure for the whole war has been found. On page 222 he gives figures of 68 employees killed and 241 injured while on company service, with another 88 killed and 255 injured while off duty.

102 Sitwell 1974, 52–3. I am very grateful to Dr Michael Turner for providing me with this quotation.

103 Nock 1967, 148–52, describes the 'Square Deal Now' campaign launched by the 'Big Four' companies in 1938.

104 Ibid, 221–3, 233–7.

105 Booker 1977, 162.

Chapter 5

1 This point is made rather more trenchantly by Nicholas Faith (1990) in the preface to his *The World the Railways Made*. There are, of course, a number of eminent exceptions to this observation, such as Barker and Robbins 1975; Kellett 1969; Simmons 1968; and Faith himself, but there have not been enough of them. E T MacDermot's great *History of the Great Western Railway*, completed by C R Clinker and O S Nock (MacDermott 1964a and b), partakes in large measure of both the strengths and weaknesses of its genre. It is a superb piece of sustained research and an invaluable resource for all students of the GWR, including the present author, but it does little to relate the GWR to the culture that created it and is fairly demanding for the general reader.

2 E C Matthews' article, 'The development of Paddington passenger station', appeared in seven parts in the *GWR Magazine* (Matthews 1916a, b and c and 1917a, b, c and d). The most valuable section for understanding Brunel's design is 1916c.

3 NA Rail 1005/11, 1st Class Express and Ordinary, 2nd Class Express and Ordinary, 3rd Class Ordinary and 3rd Class Parliamentary, single and return in each case.

4 Matthews 1916b, 251.

5 Ibid. NRNRG, GWR 13309, has designs for a 'moveable bridge' at Paddington. 18700 has a foundation plan for the departure side, undated but presumably of early 1851, which clearly shows the two open bays, 16ft (4.9m) wide, made beneath the platform for the bridges, and they are faintly indicated on the large general plan for the station, roll 6262, undated, also early 1851. One of the bridges survived into the 20th century.

6 All the drawings for machinery in NRNRG, GWR 13309 are stamped as by Sir William Armstrong & Co. For Armstrong, *see* chapter 3, note 79.

7 *Journal of the British Association*, 1854, quoted in Matthews 1916c, 251.

8 Mr John Lewis points out that the length of the traverser would govern the maximum wheelbase-length of a vehicle that could be carried on it, not the overall length, and that this could have taken any of the GWR's 4- and 6-wheel coaches of the period, a 31ft (9.5m) coach having a wheelbase of just 19ft (5.8m).

9 NRNRG, GWR 18700, now in folder 138, 'Paddington Station, plan of foundations', unsigned and undated, but clearly the foundations for Fox Henderson's first contract, so from early 1851.

10 NRNRG, GWR 18700, an annotated outline plan, again undated but clearly from early 1851, has the positions of the traversers marked in (and labelled 'traversers'); 6262, a large outline plan of the station, marks the zones with dotted lines and their breadth, 26ft (7.9m), without explicitly saying what this is.

11 NRNRG, GWR 13309 has four drawings for traversing frames, and for hydraulic machinery to power them, which seem to fit with these locations, on the departure platform. None of them explicitly shows the width of the traversers (that is, the length of the vehicle that could be got on one); the dimension of 26ft (7.9m) comes from the foundation drawing, and from the outline plan, 6262.

12 NRNRG, GWR 13309, two drawings for a 'dropping' or 'descending' platform, hydraulically powered.

13 Matthews 1916b, 253. Matthews added that 'the difficulty of the auxiliary intermediate platforms was to have been overcome by making a portion of them capable of being lowered to rail level, to permit the passage of the traverser arrangement, but so far as can be ascertained this was never effected. Probably the difficulties attendant on the installation and operation proved insurmountable.' In other words, Matthews knew about the 'descending platforms', and on the face of it, this would seem like a convincing explanation for them. However, he does not seem to have seen the drawings for them or he would have realised that their surfaces were of plain boarding and did not have rails for the traverser to run on. Equally importantly, traversers need solid foundations and, as is noted in note 9, the foundation plans for contract 1 clearly show footings for traversers but in different positions, not in the transepts and not spanning the whole width of the shed.

14 GWR 1971. Reprint of *Great Western Railway Timetable for 1865*.

15 Matthews 1916c, 251, said that the use of this area was 'not positively known', but interpreted it as being for owners of private carriages who were intending to travel in them. The area was long known to staff into relatively recent times as the 'horse arch': information from Mr Robert Thornton.

16 Ibid, 251.

17 Ibid, 253–4. Mr John Lewis has informed me that a similar standard-gauge 'railway' was installed for moving baggage at Millbay Docks, Plymouth: pers comm.

18 Ibid, 251. This is the only reference I have found to there being a royal waiting room on the arrival side. It is not clear when it was removed. The three waiting rooms had substantial brick footings, a drawing for which is in NRNRG, GWR 13314.

19 No drawing for the up parcels office has been found, but NRNRG, GWR 13315 includes an undated drawing for adding an upper floor of offices to it.

20 NRNRG, GWR 13309 has a drawing for hydraulic machinery for the sector table, dated 3 Jun 1854.

21 MacDermot 1964a, 370.

22 Ibid. 'Twelve months later it was decided to proceed with the passenger arrival

sheds and platforms, and accommodation for the locomotive establishment, which might have meant the central span of the present day station roof… .' As we know, Brunel envisaged the overall plan of the station from the outset, and there is no indication on the early sketches that he saw the central span as being for locomotives. As John Binding (2001, 53–4) demonstrates, in the original terminus at Bristol Temple Meads the engine shed was essentially the 'town end' of the main passenger shed, the equivalent to the Lawn area at Paddington.

23 MacDermot 1964a, chapter 16, 'Carriages and Wagons'. Some of the early carriages 18ft (5.5m) long, would have fitted on a 22ft (6.7m) turntable, but the 38ft (11.6m) 'Long Charlies' introduced in 1852 clearly would not.

24 MacDermot 1964a, 349–52, section on royal trains.

25 So much so that the greater part of it is today converted into a spacious shopping and eating place for tourists, with the trains confined to one mean platform at the back. *Sic transit gloria mundi.*

26 NA Rail, miscellaneous history file, 1005/66.

27 Quoted in Acworth 1889, 250.

28 MacDermot 1964a, 332–5.

29 Buchanan 2002, 164.

30 MacDermot 1964a, 339, reproduces a GWR timetable dated 30 Jul 1841.

31 Ibid, 334.

32 In 1845–6, the time of the fastest London to Bristol expresses came down to three hours or less.

33 Simmons 1968, 24–5.

34 NA Rail 267/16, General manager's report on third-class fares, 18 Jan 1874. The company had raised third-class fares experimentally in South Wales to one and a quarter old pence a mile. First-class passenger numbers had fallen by 736, second-class by 37,144, third-class numbers had risen by 1,037,808.

35 Booker 1977, 102.

36 Anon 1894.

37 Ibid.

38 MacDermot, 1964b, 235–6.

39 Simmons 1968, 145.

40 Booker 1977, 80.

41 As Mr John Lewis kindly told me, 'the first corridor train' should be expressed in the singular as the GWR had just the one set of corridor coaches for the first several months.

42 Simmons 1968, 146–7.

43 GWR 1971.

44 NA Rail 250/18, 448, GWR board meeting of 7 Jan 1863.

45 Simmons 1986, 89, quoting an article in *The Globe*, 3 Sep 1863.

46 Barker and Robbins 1975, 204.

47 NA Rail 267/166.

48 Anon 1894. At first sight, these figures look anomalous – as if each journey was of only *c* 0.3 miles (0.5km)! I think that 'passenger miles' must be interpreted as meaning passenger miles run by every train service, not by every passenger. If each train carried in the order of 50 to 300 passengers, this would seem to raise the mean journey length to a more plausible range.

49 MacDermot 1964a, 230.

50 NA Rail 267/39, report on the London horse establishment, 6 Dec 1877. With the first stage of the stables nearing completion the company had to decide whether to let them or buy their own horses. The capital cost of buying 130 horses with equipment and a year's provender and running costs was estimated at £11,168.

51 NA Rail 258/243, estimate of space allocations needed, 29 Jul 1873. This document needs to be used with care, as it is not at all clear that it is complete. It lists the following staff: accountants, 17; finance, 9; Mr Friend, 1; Mr Tyrell, 2; Mr Grant, 5; general manager, 4; board and committees, 4; chairman, 1; deputy chairman, 1; secretary, 3; messengers, 5; solicitor, 8; rates and taxes, 2; Dr Cooper, 1; estate, 3; storeroom, 1; engineers, 9. The square footage occupied by each department or person is also listed. The document is very bald and it is just possible that the numbers refer to rooms, not people, but analysis of the list suggests that this is very unlikely, nor does it seem to square with plans of the building. This only refers to the upper floors of Eastbourne Terrace, as functions like the booking office, the waiting rooms, buffet and printing press, all on the ground floor, are not listed here.

52 NA Rail 267/145. This list is stamped 3 Apr 1873, but was compiled for completely different reasons to the 'head office' list, being a table comparing the official staff complement with the staff currently in post, giving their maximum salaries, and recommending revised maximums. The official complement was: station superintendent's office, 3; booking office, 11; abstract office, 2; excess luggage, 2; cloakroom, 1; police office, 1; collector, 1; cartage, 2; down parcels, 13; up parcels, 6.

53 Ibid. The next section of this document lists 'maximum staff' at the goods depot in the same way.

54 Ibid. This document lists all of the station's office staff in all stations and, in most cases, recommends that their maximum salary be increased.

55 NA Rail 1149/7, Brunel letter books, 49. Brunel took considerable interest in the enginemen's welfare: on 18 Jan 1842 he wrote to the board listing the 16 who had given good service and qualified for the £10 bonus.

56 NA Rail 250/8, GWR board minutes 1854–5, 172.

57 Ibid, 173. It is worth pointing out that a GWR policeman was not performing a job we would associate with that title but was something more akin to a signalman. I am grateful to Mr John Lewis for pointing this out.

58 Ibid, 188.

59 Ibid, 190.

60 Ibid, 249.

61 NA Rail 267/121, report on staff and working expenses at stations, 6 Jan 1863.

62 NA Rail 250/18. The GWR board minutes for the next three meetings after 6 Jan 1863 were examined, but no reference to the report was found.

63 Information from Mr John Lewis.

64 Dodds 1953, 374. Mr Dodds does not give a source for this and I have not been able to verify it from GWR records.

65 NA Rail 1005/212, file on the GWR Literary Society.

66 *The Illustrated London News*, 17 Dec 1859.

67 The best run of the magazine is that held by STEAM, Museum of the Great Western Railway, Swindon. Other good runs are held by the National Archives (formerly Public Record Office), and by the Great Western Trust, Didcot.

68 NA Rail 253/669, catalogue of paintings etc at Paddington, 89.

69 NA Rail 1005/66, 'Miscellaneous History', document published to mark the company's 91st birthday, 1929.

70 NA Rail 1005/82, copy of Green, F W 1937. 'The working of Paddington passenger station'. *Proceedings of the GWR (London) Lecture and Debating Society*, Jan 1937.

71 Ibid. The data from here to the end of the chapter is taken from Mr Green's article.

72 Information from Mr Robert Thornton, Network Rail.

Chapter 6

1 Kadleigh, S with Horsburgh, P 1952.

2 Hitchcock 1951.

3 Information from Robert Thornton; NA Rail 1005/82, British Rail press release dated 24 Sep 1968.

4 Information from Robert Thornton. Moving the newspaper services to the new depot enabled British Rail to keep the articulated lorries, which had been loaded at night inside the station, outside.

5 Ibid. The master plan was produced in-house, the first stage of work designed by Aukett Associates.

6 Fowler 1990; Brindle 1991.

7 Information from Mr Graham King, Westminster City Council: I am grateful to Mr King for many insights relating to Paddington, its environment and planning context.

8 I am very grateful to Mr Robert Thornton of Network Rail for much information on recent works at Paddington.

9 Personal communications from Robert Thornton, and Andrew Whalley of Nicholas Grimshaw & Partners; Powell 1999.

10 Information from Graham King, Westminster City Council.

11 Alan Baxter Associates, Crossrail C131 – Paddington Integrated Project – WES/4/36/H1 – Heritage Method Statement, July 2010. I am very grateful to Graham King of Westminster City Council, to Dave Keeley of Crossrail Limited, and to Robert Thorne and Richard Pollard of Alan Baxter Associates, for much information and advice relating to Crossrail and the Paddington Integrated Project.

12 Buchanan 2002, 221–4.

Chapter 7

1 Hitchcock 1970, 559.

2 Addyman and Fawcett 1999, chapter 5.

3 Turner 1850, 204. Turner's roof was immensely influential and several mid-Victorian stations still have sickle or crescent trusses based on this model. His original Lime Street roof did not last long, being replaced by the present 212ft (64.7m) sickle truss roof in 1867.

4 Addyman and Fawcett 1999, 76.

5 NRNRG, GWR 13320.

6 The visual contribution that the 'stars and planets' make can be assessed by looking for comparison at the series of 'blank' arches in Span One (over platform 1), which were replaced after wartime damage without these decorative perforations.

7 NRNRG, GWR Collection. The GWR's MS drawings catalogue records roll 102790 as containing 39 drawings of full-size details for the ornamental cast-iron work, all signed off by Brunel. This was missing from the collection in March 1992.

8 *The Builder*, **12**, 17 Jun 1854, 322.

9 NRNRG, formerly NRWRPR microfilm 24213–7, formerly GWR 52658, drawings for new steel columns between platforms 7 and 8, Cleveland Bridge Engineering Company, 12 May 1923.

10 Grimshaw 2000.

11 BUL, Brunel Collection, office diaries: on 24 Mar 1851 Brunel met his assistant Charles Gainsford and one William Tapper about 'Paddington Station galvanised iron', meeting Gainsford and Charles Fox later the same day.

12 McKean 1994, 15; Colquhoun 2003, 166–9.

13 NA Rail 1005/82, typescript copy of the specification for Fox, Henderson's second contract, specifies 'Patent rolled plate 26 ounce [793g]' glazing for 'Paxton Glass Roofing – in part,' and no. 18 gauge galvanised corrugated-iron roofing for the lower part of the roofs.

14 NRNRG, formerly NRWRPR, microfiche 115.

15 Hitchcock 1951; Hitchcock 1972, 558–61.

16 Hitchcock 1972, 561.

17 The principal modern biography of Owen Jones by Carol Flores (2006) makes no reference to Jones' design for Paddington, and there is no reason to believe that any drawings by him relating to the station still exist.

18 *The Builder*, 12, 28 Jan 1854, 41.

19 NA Rail 1005/82, GWR half-yearly reports to shareholders.

20 NA Rail 447.

21 I have not found the original letter: it is quoted in Levien 1930b, 382.

22 NA Rail 250/187, 53. The Expenditure Committee considered 'paving approach from Spring Street and Bishop's Bridge Road with small stones'. Ibid, 56, 24 Nov 1853, the same committee advised that the inclined road to the arrival side be paved with small stones like the road by the departure platform.

23 NRNRG, GWR 13315, 13315A.

24 Andrew Harris, Crossrail project, document no. C130-SWN, 'Paddington Station, City of Westminster Heritage Agreement, Heritage Method Statement, Departures Road Canopies & Eastbourne Terrace Retaining Wall & Railings', WES/4/5/H5, July 2010. Under the Crossrail project, the surviving run of 1850s railings is to be dismantled and retained for reuse along Eastbourne Terrace.

25 Readers who try counting the bays may find themselves foxed: the slightly projecting end sections of the Eastbourne Terrace façade had six bays at upper level but only seven bays at street level. The left-hand section with the original entrance to the royal waiting room, survives: the right-hand section was demolished after wartime bomb damage.

26 NRNRG, GWR 13315. The elevation is in a roll of drawings of full-size details for the external treatment of the façade.

27 NRNRG, GWR 13328 contains a fine drawing for the ceilings of the vestibule, newspaper and refreshment rooms.

28 NRNRG, GWR 17021, contract drawings signed by Lancaster Owen and dated May 1876.

29 NRNRG, formerly NRWRPR microfiche 118 and 119, plan and details of Paxton roofing.

30 NRRC, GWR 17844 has the original drawings for this building, initialled by E O (probably E Olander) for the GWR on 20 Oct 1880, and signed by Lancaster Owen for the GWR and the contractors Kirk & Randall of Woolwich on 28 Nov 1880. It is not clear who designed it, though either Olander or Owen are a possibility. NA Rail 258/343 has a detailed specification for the building.

31 NRNRG, GWR 63400, 33 drawings for new departure side offices dated 4 May 1931; 63588, 6 steelwork drawings.

32 NRNRG, GWR 64611, 14 architectural drawings and 127 steelwork drawings for a further extension of the departure side offices.

33 NRNRG GWR collection has numerous sets of drawings for the 1880s office expansion, which evidently took place in phases. They are somewhat difficult to interpret. Roll 17032, drawings for a proposed new storey over existing offices; roll 17153 has useful general plans of the offices with their uses marked on, as well as plans for 'additional offices' referring to the south-eastwards extension; roll 17840, 'fourth additional offices' refers to the south-east extension; roll 17841, 'additional offices', again refers to the south-east extension and the additional storeys, which were evidently handled under the same contract, by Messrs Kirk & Randall of Woolwich.

34 NRNRG, GWR 17162, 'Additional Storey of Office'; 17841, 'Girders over Booking Hall'.

35 NA Rail 1008/64, misfiled letter from Brunel to Saunders, 23 Jan 1851.

36 NRNRG, formerly NRWRPR, microfiche 118, 125.

37 NRNRG, GWR 13315A contains numerous coloured full-size drawings for these.

38 There are drawings for the extension of the cab roof in two stages, 1880–1. NRNRG, GWR 17840, drawing no. 30, '4th additional offices – extension of Paxton Roofing', signed by Lancaster Owen and Kirk & Randall, 23 Nov 1880, brings the cab roof from the south-east end of Brunel's original 32-bay centre, for another eight bays, to the edge of the new block, and indicates that the 'present fascia' was to be repaired and re-erected at the new south-east end. NRNRG, GWR 17841 (microfiche 33767), also '4th

additional offices, extension of Paxton Roofing', initialled E O and dated 2 Apr 1881, has the roofing brought another seven bays eastwards, across seven of the nine bays of the new block. The details of the canopy ironwork are slightly different between the two drawings.

39 Andrew Harris, Crossrail project, document no. C130-SWN, 'Paddington Station, City of Westminster Heritage Agreement, Heritage Method Statement, Departures Road Canopies & Eastbourne Terrace Retaining Wall & Railings', WES/4/5/H5, Jul 2010.

40 NA Rail 250/5, GWR board minutes, 262, 4 Dec 1851. Brunel submitted plans for the board and committee rooms and general offices, above the booking offices on the departure side. Surviving drawings of these are held in NRNRG, GWR 13315, 13328 (dated 1851–2) and 16144 (dated 1853–5).

41 NRNRG, GWR 13328, roll of full-size details for interior ornament, including the cornices and ceiling details for the vestibule, newspaper and refreshment rooms, the first-class waiting room and lobby, for the committee room and Mr Saunders' room.

42 NRNRG, GWR 13315, undated drawing for the lantern.

43 NRNRG, GWR 13607–13626 (formerly 52629).

44 See chapter 3, note 88.

45 NRNRG, GWR 13328, roll of full-size details of ornament for interiors; 16144, designs for offices and fittings, 1855.

46 NA Rail 258/243, undated memorandum, filed with documents of c 1916.

47 NA Rail 253/669, D V Levien's catalogue of paintings and works of art at Paddington in 1926 lists well over a thousand objects. The version of Frith's painting was one of the copies made by Marcus Stone: see chapter 4 notes 2 and 3, and was a gift from Frank Bibby, a director, in 1917. The engravings were a gift from Sir Watkin Williams Wynn. The seals remain in the collection of STEAM, Museum of the Great Western Railway, Swindon.

48 Levien 1908 and 1909. Some of the objects are now in the National Railway Museum, York, and STEAM at Swindon.

49 I am very grateful to Mr Richard Pollard of Alan Baxter Associates for telling me about this discovery.

50 Matthews 1916a, b and c.

51 Matthews 1917a.

52 NRNRG, formerly NRWRPR microfilm 1474–1522 (the original GWR numbers are not legible on the microfilm, and the drawings are not visibly dated. I am very grateful to Mr Dave Keeley of Crossrail for

making this material (and much more) available to me; Matthews 1917a, b, c, and d; Andrew Harris, 'Crossrail Multi-Disciplinary Consultant Works, Package 2, Paddington Station: Building Recording at the London Street Deck' document no. CR-DV-PAD-X-RT-000062, 2010.

53 Matthews 1917d.

54 NRNRG, GWR 59866, details of the new roof. A large-scale drawing of 'Details of the New Roof' shows the five centres with their radii. It shows the 'stars and planets', and the lower panels of cast-iron tracery at the springing of the arches, but not the smaller, upper sections of tracery that had figured in Brunel and Wyatt's original design: perhaps they had been removed from the rest of the roof by this time.

55 Matthews 1917d; NRNRG, formerly NRWRPR, microfilm 29827–29840 formerly GWR 51526, drawings for roof over platforms 9–11, contract 12, 1913; microfilm 28719–28739, formerly GWR 52131, old no. 59968, drawings for contract 14, Apr 1914; 'Steel Structures', Apr 1919, 108–12; NA Rail 267/304, report on Paddington Station roof, 9 July 1915.

56 Shackle 1933, 403–4, 410–11.

57 NRNRG, GWR 63776, drawings of details of steelwork in the deck and strengthening to the ramp, with plan and elevation of the new building dated April 1932 to February 1933.

58 Information from Robert Thornton and Nick Hartnell, Network Rail.

59 'SAVE Paddington's Span Four – This Engineering Marvel Must Stay', Report by SAVE Britain's Heritage, London, 2006.

60 I am very grateful to Graham King of Westminster City Council, Dave Keeley of Crossrail Ltd, and to Robert Thorne and Richard Pollard of Alan Baxter Associates, for information about the Crossrail project and its impact on Paddington.

61 NRNRG, formerly NRWRPR microfilm 1344–61, formerly GWR 63687, 1932; Culverhouse 1934.

62 Culverhouse 1934.

63 Anon 1933a.

64 The most important contender for the title of first grand railway hotel is the North Euston at Fleetwood, designed by Decimus Burton as part of Sir Peter Fleetwood-Hesketh's development of the town, c 1841–7. The scheme as a whole depended on the assumption that the London & North Western would take its Glasgow services from Euston to Fleetwood (hence the hotel's name) and thence by ferry to the Clyde. This was wrecked by Joseph Locke's success in taking the Lancaster to Carlisle line over Shap Summit in 1843–6. Sir Peter was

bankrupted. Burton had envisaged the North Euston as a giant circular building, of which about a third was built. It remains in hotel use.

65 Hitchcock 1972, 211–12.

66 Lever 1973, 88.

67 Obituary of John Thomas, The Builder, 20, 19 Apr 1862, 275. The Illustrated London News devoted an article to the pediment on 2 Sep 1854, 217.

68 The Illustrated London News, 18 Dec 1852.

69 Hardwick's perspective of the coffee room is in the Royal Institute of British Architects Drawings Collection, see Lever 1973, 88.

Chapter 8

1 The Builder, 21, 10 Jan 1863, 21–3.

2 Anon 1914.

3 NRNRG, GWR 7494, undated plan for the 'London Depot', and another undated plan marked 'GWR London Terminus' show variants of this plan.

4 NRNRG, GWR 14909–21, fine series of coloured drawings for the first goods depot.

5 BUL, Brunel Collection, Large Sketchbook 4, 15, 29 Jan 1851.

6 NA Rail, 250/5, 53, 73, 98.

7 NRNRG, GWR 13324, roll of 17 drawings for the goods station, one dated 17 Oct 1851.

8 NRNRG, GWR 17842, general plan of the goods yard; 30658 includes an isolated drawing for 'Roofing to Coal Viaduct' dated Aug 1906; NA Rail 250/49, 230.

9 NA Rail 250/6, 381–2, GWR board minutes for 10 Mar 1853.

10 NRNRG, GWR 16122, roll of 22 drawings, for warehouse and goods sheds; 13304 has a number of isolated drawings in a larger roll including a design for a bowstring truss, tracing sent to Gainsford on 23 Jun 1854, and to Bertram on 24 Feb 1856; details of the warehouse roof dated 30 Aug 1855; 17842, roll of drawings for 'Contract G', relating to the second phase of the goods depot.

11 NRNRG, GWR 13309, drawing of lifting platforms for goods shed, 1853; 13312, 16 drawings of hydraulic machinery, including hydraulic accumulator towers, capstans and gearing, hoists and cranes for the goods shed, several of them signed as by Sir William Armstrong & Co, Oct 1856; Brunel 1870, 85, note 1, referring to an article by Armstrong in the Report of the British Association, 1854, 418.

12 NRNRG, GWR 17843, plans for 'new goods offices', carried over the far (east) end of the tracks, within the first, northern part of the goods depot.

13   *The Railway Engineer*, Mar and Apr 1906, has articles on the completion of the building: information from Mr Peter Rance.

14   Bond 1925.

15   NRNRG, GWR 49753, new roof to the Birmingham platform, Dec 1908; 50525, 3 drawings for new platforms coverings, 1911; 50533, alterations to north-west corner of goods shed, undated; 50515, roof covering, extension of platform 1, Mar 1912; 52152, reconstruction of roof to one bay of goods station, Oct 1915.

16   NRNRG, the manuscript catalogue of GWR drawings lists: 52666, 7 contract drawings for goods shed roof, Apr 1925, J C Lloyd; 52667, 10 drawings, goods station alterations, Lloyd; 62034, goods depot, reinforced concrete cellarage, 1925, Indented Bar & Concrete Engineering Co; 62307, goods depot alterations, and approach to high-level yard, May 1927; 62311, goods station cellarage, ferroconcrete drawings by the Indented Bar & Concrete Engineering Co, 1927; 62393, goods shed roof, 45 contract drawings, 1927, Lloyd. Many of the drawings listed here are apparently no longer in the collection, and may have been destroyed.

17   Bond 1925.

18   I am grateful to Mr Peter Rance of the Great Western Trust for this information. The Trust hope to put these objects on display in a new GWR museum and reference library at Didcot as and when funding is available.

19   NA Rail 250/6, 71. NRNRG GWR 13306, a roll of drawings headed 'Paddington–Narrow Gauge Engine House', in fact seems mostly to consist of copies of the contract drawings for the broad-gauge engine house. It had a central section, 462ft (140.9m) long and 68ft (20.7m) wide inside, with slightly narrow extensions about 100ft (30m) long and 56ft (17m) wide inside at either end. The shed could comfortably have housed 50 locomotives and tenders (though it would have taken a long time to have got the ones in the middle out), and it is not clear why Brunel wanted so vast a building.

20   NRNRG GWR 13306, drawings for 'Paddington Locomotive Offices'. Two other rolls have copies of what appears to be an earlier version of the design, with 10 bays of workshops framed by symmetrical two-storey wings with bay windows, with the engine-drivers' accommodation at one end and Gooch at the other: 13311, a roll of drawings for carriage sheds and workshops includes an undated plan and elevation of this design: 16137 includes a drawing for a wrought-iron beam in

the 'floor to Mr Gooch's Office' dated 14 Oct 1853; 13314, a roll of miscellaneous drawings, another undated version of this earlier design.

21   NRNRG, GWR 13306 has a plan and elevation marked 'Narrow Gauge Engine House' showing it apparently as built, 12 bays long. Curiously, the roll contains an alternative version showing exactly the same design but extended to 27 bays, giving it internal dimensions of 327ft (99.7m) by 42ft (12.8m). The Ordnance Survey 25in. map confirms that it was the smaller 12-bay version which was built.

22   White 1906.

23   I am very grateful to Dave Keeley and Jay Carver of Crossrail Ltd for arranging access to the Old Oak Common site. Mr Keeley supplied a large amount of primary information on the history of Old Oak Common in digital form, notably a large number of scanned drawings from Network Rail's collection. The references to drawings of Old Oak Common, given later, are to these scans rather than to the original drawings. The history of Old Oak Common given here relies heavily on an archaeological and historical assessment of the site commissioned for the Crossrail project: Thompson, Gould and Mazurkiewicz 2010 (unpub report). I am grateful to Crossrail Ltd for supplying me with this excellent document and permission to make use of it. I have also made use of an independent report by Tucker 2010 (unpub report).

24   Tuplin 1958; Obituary notice, 'George Jackson Churchward', *Proceedings of the Institute of Mechanical Engineers*, **125**, 1933, 783; Marshall 2003.

25   Thompson, Gould and Mazurkiewicz 2010 (unpub report), 19–20, quoting TNA RAIL 250/334, 17, 148; 250/335, 150; 250/379/66.

26   Ibid, 21, quoting TNA RAIL 250/339, 255–7; 250/340, 145–6; 250/70, 97.

27   Ibid, 22, quoting TNA RAIL 250/340, 253; 250/46, 11–12.

28   White 1906; Thompson, Gould and Mazurkiewicz 2010 (unpub report), 22–4, quoting TNA RAIL 250/271, 124; 250/47, 52, 86; 252/1340.

29   Drawings for the carriage sidings, NRNRG, GWR 104300; carriage shed, NRNRG, GWR 90097, 30363, 34885; general arrangement, 137769.

30   NRNRG, GWR 82725, plan of marshalling yard.

31   NRNRG, GWR 112575, general layout of engine shed, 106195; White 1906.

32   NRNRG, GWR 43193, boiler washing plant, Dec 1910; 83887–8, refuse destructor, 1927.

33   Thompson, Gould and Mazurkiewicz 2010

(unpub report), 27.

34   Ibid, 27–32.

35   Ibid, 32; NRNRG, GWR 83887–8, Mar 1949, 127624, Feb 1952.

36   Thompson, Gould and Mazurkiewicz 2010 (unpub report), 32–4.

37   Ibid, 34; NRNRG, GWR 150294B.

38   Ibid, 34–5.

39   Tucker 2010 (unpub report). NRNRG, GWR 89250, drawing for overhead crane; 138962, drawing for replacement of traverser, 1958.

40   NRNRG, GWR 21396–8, sand furnace, 1902; 108769B, conversion to mess room, 1937.

41   A drawing was found by the Crossrail project's research for this truss design for the main engine shed at Old Oak Common, but no GWR number is legible on the scan. The Polonceau truss was invented in 1839 by the French engineer Camille Polonceau (1813–59).

42   Thompson, Gould and Mazurkiwciez, 2010 (unpub report), 27–8; NRNRG, GWR 105183, 112083, 112172, 112175, 112295, drawings of carriage paint shop and lifting shop, 1938.

43   Tucker 2010 (unpub report); Lyons 1972.

44   NRNRG, GWR 13311, undated general plan of carriage department sheds and workshops; 10266, now folder HC 189, contract drawings for the carriage department's sheds, offices, stores and workshops, dated between Aug 1855 and Jan 1856.

45   NRNRG, GWR 13311, undated plan, elevations and roof details for carriage department stores and offices.

46   NRNRG, GWR 13305, plan for smiths' and fitters' shops, signed J W for C S, 20 Dec 1853.

47   NRNRG, GWR 13305, undated, coloured contract drawings for Paddington wagon shops.

48   Anon 1907a and b. Information on recent developments from Graham King, Westminster City Council.

49   NRNRG, formerly NRWRPR microfilm 29841–6, previously GWR 63121, foundations and preparatory works on the site between Westbourne Terrace and Bishop's Road, R Carpmael, 1930.

50   NRNRG, formerly NRWRPR microfilm 1306–1343, previously GWR 63506, full contract drawings dated Jan to Jul 1932.

51   NRNRG, GWR 16144, an unlabelled, undated plan of the ground and first floors of a rectangular building, is probably of the staff hostel.

52   Anon 1933b.

53   Smith 1993 (unpub report); Thorne 1990 (unpub report).

54   NRNRG, GWR 10242–6, contract drawings for the first phase of work, 1876;

17029, plan showing the southern part of the site with the first phase built, and the whole intended layout, with a contract specification signed by Owen, Oct 1876.

55 Smith 1993 (unpub report); Thorne 1990 (unpub report); Gordon 1893, 50–3; Collier 1909.

56 Matthews 1917a, 30–1. NRNRG microfilm 3758–77, formerly 50523, 23 drawings for additions to the Mint stables by W Armstrong, GWR new-works engineer and his office, including drawings for reinforced concrete work by L G Mouchel & Partners.

57 Matthews 1917a, 30–1; NRWRG, formerly NRWRPR, microfilm 3758–77.

58 Smith 1993 (unpub report); Thorne 1990 (unpub report).

59 Gordon 1893.

60 Harwood, E 'Paddington British Rail maintenance depot, No 179 Harrow Road, City of Westminster', (unpublished historical report), Historic Analysis and Research Team, English Heritage.

61 Killick 1969, 151–7.

62 Nicholas 1914. The girders carrying the south span, 15ft (47m) long and 17ft (5.2m) deep at the centre, were reckoned to be, for their length, the heaviest ever commissioned by the GWR: the eastern one was calculated to carry 230 tons (234 tonnes) more than its counterpart.

63 NA Rail 1149/44, 'GWR Tenders'.

64 Anon 1909; NA Rail 250/49, £403 to E C and J Keay for ornamental cast-iron work for the Ranelagh Bridge.

65 No early views or drawings of the bridge have come to light, but NA Rail 1149/44 'GWR Tenders' gives the date of the contract, and an outline schedule of prices, for brickwork, concrete and stonework.

66 NA Rail 250/49, GWR board minutes, 93.

67 Anon 1910.

68 NA Rail 250/48, GWR board minutes, 147, 173. NRNRG, GWR 30658, miscellaneous drawings, includes an undated, early 20th-century drawing for widening the bridge carrying the Bishop's Road over the canal, in two spans, of 35ft (10.7m) and 16ft (4.9m).

69 Anon 1908.

70 Personal communication from Graham King, Westminster City Council, and an online report by Dorman Long Technology: 'Bishops Bridge replacement, Paddington Station Long Term Vehicle Access project, UK', http://www.dormanlongtechnology.com/ en/projects/Bishops%20bridge.htm

71 Brindle and Tucker 2004.

72 Brindle and Tucker 2008; Brindle and Tucker forthcoming.

73 Brindle and Tucker 2004.

74 NA Rail 1149/4, Brunel, letter book III, 1838, 82–3.

75 NA Rail 1149/4, 95–6.

76 NA Rail 830/6. Board minutes of the Grand Junction Canal Company, 1835–8, 293–4.

77 87 NA Rail 1149/44, 'GWR Tenders', 35, 'London Terminus – various works'.

78 NA Rail 1149/8–13.

79 NA Rail 1149/9, 'Facts', 1836–40, 96–9.

80 Brindle and Tucker 2004; Brindle and Tucker forthcoming.

81 BUL Brunel Collection, 'Miscellaneous Sketchbook', 68–70.

82 Institution of Civil Engineers Library: I am very grateful to Mr Mike Chrimes for recognising the drawings (which are neither labelled, signed nor dated) and telling me about them, shortly after the bridge's discovery. The drawings were purchased by the ICE with a body of material probably produced by Francis Trevithick, including several drawings for works on the London end of the GWR main line: it would seem likely that Trevithick was making or copying drawings for Brunel on a freelance basis, Brunel's own drawing office at Duke Street presumably being at full stretch.

83 Brindle and Tucker 2008; Brindle and Tucker forthcoming.

84 Brindle and Tucker forthcoming.

85 Reference to press stories, Brunel Bridge, 4 Mar 2004: *The Guardian*; *The Times*, 8–9; *The Daily Telegraph*, 13; *Daily Mirror*, 20.

Span Four as restored, photographed in 2012. [© English Heritage DP157917]

BIBLIOGRAPHY

Acworth, W M 1889 *The Railways of England*. London: John Murray

Addyman, J and Fawcett, B 1999 *The High Level Bridge and Newcastle Central Station: 150 Years across the Tyne*. Newcastle upon Tyne: North Eastern Railway Association

Anon 1894 'An interview with the General Manager'. *GWR Magazine*, Nov 1894, 3–4

Anon 1907a 'The engineering department – New stationery stores, etc, Paddington'. *GWR Magazine*, Mar 1907, 63–4

Anon 1907b 'Reinforced concrete in railway structures'. *GWR Magazine*, Nov 1907, 243–5

Anon 1908 'The engineering department – Bishop's Road Bridge'. *GWR Magazine*, Feb 1908, 39–40

Anon 1909 'The engineering department – Ranelagh Bridge, Paddington'. *GWR Magazine,* Jul 1909, 159

Anon 1910 'The engineering department – reconstruction of Westbourne Terrace Bridge'. *GWR Magazine*, Aug 1910, 207–9

Anon 1914 'Departmental news – connection of Paddington Station with the "Bakerloo" railway'. *GWR Magazine*, Jan 1914, 17

Anon 1933a 'New offices at Paddington Station'. *GWR Magazine*, Mar 1933, 110–11

Anon 1933b 'Structural alterations at Paddington – New offices and store for stationery department and hostel for refreshment department staff'. *GWR Magazine*, Sep 1933, 374

Anon 1933c 'Modernising Paddington Station, GWR'. *Railway Gazette*, 8 Sep 1933, 345–52

Anon 1933d 'Paddington modernised, Great Western Railway'. *The Railway Magazine*, Dec 1933, 419–26

Anon 1935 'Current notes and news – A quick lunch and snack bar at Paddington Station'. *GWR Magazine*, Aug 1935, 532–3

Bailey, M (ed) 2003 *Robert Stephenson – the Eminent Engineer*. Aldershot: Ashgate Publishing

Barker, T C and Robbins, M 1975 *A History of London Transport. Vol 1: The Nineteenth Century*, rev edn. London: Allen & Unwin

Beattie, S 1975 *Royal Institute of British Architects Drawings Collection Catalogue: Alfred Stevens*. Farnborough: Gregg International

Biddle, G 1986 *Great Railway Stations of Britain: Their Architecture, Growth and Development*. Newton Abbot: David & Charles

Bills, M and Knight, V (eds) 2007 *William Powell Frith – Painting the Victorian Age*. London and New Haven: Yale University Press

Binding, J 2000 'The final bridge: The design of the Royal Albert Bridge, Saltash' *in* Kentley, E, Hudson, A and Peto, J (eds), *Isambard Kingdom Brunel: Recent Works*. London: The Design Museum, 86–99

Binding, J 2001 *Brunel's Bristol Temple Meads: A Study of the Design and Construction of the Original Railway Station at Bristol Temple Meads 1835–1965*. Hersham: Oxford Publishing

Bond, A 1925 'Reconstruction of Paddington goods station'. *GWR Magazine*, Sep 1925, 343–6

Booker, F 1977 *The Great Western Railway, A New History*. Newton Abbot: David & Charles

Bourne, J C 1839 *Drawings of the London & Birmingham Railway*. London: Ackerman

Bourne, J C 1846 *The History and Description of the Great Western Railway*. London: David Bogue

Brindle, S 1991 'Paddington Station Roof '. *English Heritage Conservation Bulletin*, Jun 1991, 19–20

Brindle, S 2005 *Brunel – The Man who Built the World*. London: Weidenfeld & Nicolson

Brindle, S and Tucker, M 2004 'Brunel's lost bridge: The rediscovery and salvage of the Bishops Road Canal Bridge, Paddington'. *Construction History* **20**, 45–70

Brindle, S and Tucker, M 2008 'I K Brunel's first cast iron bridges and the Uxbridge Road fiasco'. *Trans Newcomen Soc* **78**, No. 1, 25–45

Brindle, S with Tucker, M forthcoming *Brunel's Cast Iron Bridges: A Descriptive and Analytical Catalogue*. Digital publication. English Heritage

Brunel, I 1870 *The Life of Isambard Kingdom Brunel, Civil Engineer*. London: Longmans, Green (reprinted 1970, Newton Abbott: David & Charles)

Bryan, T 2000 'Brunel's broad gauge' *in* Kentley, E, Hudson, A and Peto, J (eds), *Isambard Kingdom Brunel: Recent Works*. London: The Design Museum, 37–52

Buchanan, R A 2000 'Working for the chief: The design team and office staff of I K Brunel' *in* Kentley, E, Hudson, A and Peto, J (eds), *Isambard Kingdom Brunel: Recent Works*. London: The Design Museum, 14–24

Buchanan, R A 2002 *Isambard Kingdom Brunel*. London: Hambledon & London

Cattell, J and Falconer, K 1995 *Swindon: The Legacy of a Railway Town*. London: HMSO

Chapel, J 1982 *Victorian Taste: The Complete Catalogue of Paintings at the Royal Holloway College*. London: Zwemmer

Collier, E J 1909 'Railway horses'. *GWR Magazine*, Dec 1909, 279–81

Colquhoun, K 2003 *A Thing in Disguise: The Visionary Life of Joseph Paxton*. London: Harper

Conder, F R 1983 *The Men Who Built Railways*. (First pub 1868 as *Personal Recollections of English Engineers*, London: Hodder and Stoughton) 1983 edn, Simmons, J (ed), London: Thomas Telford

Cowper, E A 1854 'Description of the wrought-iron roof over the central railway station at Birmingham'. *Proc Inst Mechanical Engineers*, 26 Jul 1854, 79–87

Cross-Rudkin, P and Chrimes, M (eds) 2008 *Biographical Dictionary of Civil Engineers, Vol 2, 1830–1890*. London: Institution of Civil Engineers

Culverhouse, P E 1934 'The alterations at Paddington Station: Part iv, the new buildings'. *GWR Magazine*, Feb 1934, 65–83

Cusack, P 1986 'Architects and the reinforced concrete specialist in Britain 1905–08'. *Architect Hist* **29**, 183–95

Cusack, P 1987 'Agents of change: Hennebique, Mouchel and ferroconcrete in Britain, 1897–1908'. *Construction History* **3**, 61–74

Dixon, R and Muthesius, S 1978 *Victorian Architecture*. London: Thames & Hudson

Dodds, J W 1953 *The Age of Paradox: A Biography of England, 1841–51*. London: Gollancz

Ellaway, K J 1994 *The Great British Railway Station: Euston*. Oldham: Irwell

Ellis, C H 1959 *British Railway History: An Outline from the Accession of William IV to the Nationalisation of the Railways. Vol 2: 1877–1947*. London: Allen & Unwin

Ellis, C H 1975 *The Royal Trains*. London: Routledge & Kegan Paul

Evans, D M 1859 *The History of the Commercial Crisis, 1857–8*. London: Groombridge

Faith, N 1990 *The World the Railways Made*. London: Bodley Head

Flores, C 2006 *Owen Jones, Design, Ornament, Architecture and Theory in an Age of Transition*. New York: Rizzoli

Fowler, D 1990 'Paddington bares all'. *New Civil Engineer*, 15 Mar 1990, 40–1

Foyle, A 2004 *Pevsner Architectural Guides: Bristol*. London and New Haven: Yale University Press

Gordon, W H 1893 *The Horse World of London*. London (reprinted 1971, London: J A Allen)

Griffiths, D 2000 'The Leviathan: Designing the *Great Eastern*', in Kentley, E, Hudson, A and Peto, J (eds), *Isambard Kingdom Brunel: Recent Works*. London: The Design Museum, 112–27

Grimshaw, N 2000 'Paddington station – project assessment', in Kentley, E, Hudson, A and Peto, J (eds), *Isambard Kingdom Brunel: Recent Works*. London: The Design Museum, 69–70

GWR 1971 *Great Western Railway Timetable for 1865*. Reprinted edn, Oxford: Oxford Publishing

Hitchcock, H-R 1951 'Brunel and Paddington'. *The Architectural Review* **109**, Apr 1951, 240–6

Hitchcock, H-R 1972 *Early Victorian Architecture in Britain*. (1st edn 1954) 1972 edn, London: Trewin & Copplestone

Hobhouse, H 1976 'P and P C Hardwick: An architectural dynasty' in Fawcett, J (ed) *Seven Victorian Architects*. London: Thames & Hudson, 32–49

Jackson, A 1969 *London's Termini*. Newton Abbott: David & Charles

Jones, O 1856 *Grammar of Ornament*. London: Day and son

Kadleigh, S with Horsburgh, P, 1952 *High Paddington, a Town for 8,000 People*. London: The Architect and Building News

Kellett, J R 1969 *The Impact of Railways on Victorian Cities*. London: Routledge & Kegan Paul

Kentley, E, Hudson, A and Peto, J (eds) 2000 *Isambard Kingdom Brunel: Recent Works*. London: The Design Museum

Killick, J 1969 'Paddington maintenance depot'. *Official Architecture and Planning*, Feb 1969, 151–7

Latimer, J, 1887 *The Annals of Bristol in the Nineteenth Century*. Bristol: W & F Morgan

Lever, J (ed) 1973 *Catalogue of the Drawings Collection of the Royal Institute of British Architects: G–K*. Farnborough: Gregg International

Levien, D V 1908 'Pictures at Paddington, No 1'. *GWR Magazine*, Jul 1908, 147–50

Levien, D V 1909 'Pictures at Paddington, No 2'. *GWR Magazine*, Aug 1909, 177–9

Levien, D V 1930a 'Through the Great Western Museum, No 2 – Frith's famous picture, "The Railway Station" '. *GWR Magazine*, Mar 1930, 109–11

Levien, D V 1930b 'Through the Great Western Museum, No 3 – Paddington and Paddington Station '. *GWR Magazine*, Sep 1930, 379–82

Levien, D V 1936 'The hitherto untold story of "Punch's Railway" – an account of the unfortunate history of the West London Railway Company'. *GWR Magazine*, May 1936, 221–4; Jun 1936, 272–6; Jul 1936, 325–8; Sept 1936, 441–4

E T Lyons, 1972 *An Historical Survey of Great Western Engine Sheds, 1947*. Oxford: Oxford Publishing Company

MacDermot, E T 1964a *History of the Great Western Railway. Vol I: 1833–63*. (1st edn 1927) 1964 rev edn, Clinker C R (ed), London: Ian Allan

MacDermot, E T 1964b *History of the Great Western Railway. Vol II: 1863–1921*. (1st edn 1927) 1964 rev edn, Clinker C R (ed), London: Ian Allan

Marshall, J 2003. *A Biographical Dictionary of Railway Engineers*. 2nd edn, Oxford: Railway & Canal Historical Society

Matthews, E C 1916a 'The development of Paddington passenger station'. *GWR Magazine*, Jul 1916, 159–62

Matthews, E C 1916b 'The development of Paddington passenger station, II'. *GWR Magazine*, Sep 1916, 203–5

Matthews, E C 1916c 'The development of Paddington passenger station, III'. *GWR Magazine*, Nov 1916, 251–4

Matthews, E C 1917a 'The development of Paddington station, IV'. *GWR Magazine*, Feb 1917, 28–31

Matthews, E C 1917b 'The development of Paddington station, V'. *GWR Magazine*, Apr 1917, 69–70

Matthews, E C 1917c 'The development of Paddington station, VI'. *GWR Magazine*, Sep 1917, 175–7

Matthews, E C 1917d 'The development of Paddington station, VII'. *GWR Magazine*, Dec 1917, 236–8

McKean, J 1994 *The Crystal Palace: Joseph Paxton and Charles Fox*. London: Phaidon

Meeks, C L V 1957 *The Railway Station: An Architectural History*. London: Architectural Press

Menear, L 1983 *London's Underground Stations: A Social and Architectural History*. Tunbridge Wells: Midas

Mitchell, V and Smith, K 2000 *Western Main Lines: Paddington to Ealing*. Midhurst: Middleton Press

Nicholas, J S 1914 'The reconstruction of Lord Hill's Bridge at Royal Oak Station, (Paddington)'. *GWR Magazine*, Feb 1914, 37–41

Nock, O S 1962 *The Great Western Railway in the 19th Century*. London: Ian Allan

Nock, O S 1967 *History of the Great Western Railway. Vol III, 1923–1947*. London: Ian Allan

Owings, N A 1973 *The Spaces in Between: An Architect's Journey*. Boston, Mass: Houghton Mifflin

Pevsner, N 1950 *Matthew Digby Wyatt*. Cambridge University Press, Cambridge

Pevsner, N 1976 *A History of Building Types*. London: Thames & Hudson

Pevsner, N 1978 *The Buildings of England: Derbyshire*. 2nd edn, revised Williamson, E. Harmondsworth: Penguin

Pole, F J C 1926 *Great Western Railway, General Strike*. Paddington: Great Western Railway

Powell, K 1999 'Station masters'. *Architects' Journal*, 30 Sep, 24–33

Reed, M C 1996 *The London and North Western Railway: A History*. Penryn: Atlantic Transport

Rolt, L T C 1970 *Isambard Kingdom Brunel*. (1st edn 1957) 1970 edn. Harmondsworth: Penguin

Ruskin, J 1849 *Seven Lamps of Architecture*. London: Smith, Elder & Co

Saint, A J, Pettit, S and Irlam, G 1992 *Cragside*. London: National Trust

Shackle, C E 1933 'The structural alterations at Paddington Station'. *GWR Magazine*, Oct 1933, 399–417

Simmons, J 1968 *The Railways of Britain*. London: Macmillan, Oct 1933, 399417; Nov 1933, 449–67

Simmons, J 1986 *The Railway in Town and Country*. Newton Abbot: David & Charles

Simmons, J (ed) 1971 *The Birth of the Great Western Railway: Extracts from the Diary and Correspondence of George Henry Gibbs*. Bath: Adams & Dart

Sitwell, O 1974 *Queen Mary and Others*. London: Michael Joseph

Stretton, C 1901, *The History of the Grand Junction Railway*. Leeds: Goodall and Suddick

Sutherland, R J M 1975 'Oxford Midland Station and the Crystal Palace'. *The Structural Engineer* 53, 69–72

Sutherland, R J M 1989 'Shipbuilding and the long span roof '. *Trans Newcomen Soc* **60**, 107–26

Thorne, R 1978 *Liverpool Street Station*. London: Academy Editions

Thorne, R 1985 'Masters of building: Paddington Station'. *Architects' Journal* **182**, 44–58

Toppin, D 1981 'The British hospital at Renkioi, 1855'. *The Arup Journal* **16** (2), 2–18

Trench, R and Hillman, E 1984 *London under London*. London: John Murray

Tuplin, W A 1958 *Great Western Steam*. London: Allen & Unwin

Turner, R 1850 'A description of the iron roof over the railway station, Lime-street, Liverpool'. *Proc Inst Civil Engineers* **19**, 1850, 204

Tutton, M 1999 *Paddington Station, 1833–54: A Study of the Procurement of Land for, and Construction of, the First London Terminus of the Great Western Railway*. Mold: Railway and Canal History Society

Tweedie, M G 1906 'Gordon electric light installation at Paddington'. *GWR Magazine*, Dec 1906, 242–5

Vaughan, A 1991 *Isambard Kingdom Brunel: Engineering Knight Errant*. London: John Murray

Vaughan, A 1997 *Railwaymen, Politics and Money: The Great Age of Railways in Britain*. London: John Murray

Victoria County History 1976 *A History of the County of Stafford. Vol XVII*. Oxford: Oxford University Press

Waters, L 1993 *London: the Great Western Lines*. Shepperton: Ian Allan

White, A J L 1906 'New locomotive depot at Old Oak Common'. *GWR Magazine*, May 1906, 85–9

White, H P 1987 *A Regional History of the Railways of Great Britain. Vol III: Greater London*. 3rd rev edn, Newton Abbott: David & Charles

Wilson, R B (ed) 1972 *Memoirs and Diary of Sir Daniel Gooch, Transcribed from the Original Manuscript and Edited with an Introduction and Notes by Roger Burdett Wilson*. Newton Abbot: David & Charles

Wodehouse, P G 1954 *Uncle Fred in the Springtime*. (1st edn 1939) 1954 edn, Harmondsworth: Penguin

Wood, C 2006 *William Powell Frith: A Painter and his World*. Stroud: Sutton Publishing

Wyatt, M D 1850 'Iron work and the principles of its treatment'. *Journal of Design* **4**, 77–8

Unpublished reports

Richardson, S 1999 'Paddington Basin, London W2: An archaeological addendum to the archaeological impact assessment'. AOC Archaeology

Smith, J 1993 'Historic buildings report for the Mint Wing, St Mary's Hospital, Marylebone, the former Mint Stables, South Wharf Road, Westminster'. (Royal Commission on the Historical Monuments of England) English Heritage Archive File BF090837

Thompson, G, Gould, M and Mazurkiewicz T 2010 'Crossrail central section project: Old Oak Common worksites, archaeological detailed desk-based assessment: Non-listed built heritage'. Document no. C150-CSY-T1-RGN-CR076_PT001–00011, PreConstruct Archaeology

Thorne, R 1990 'The Mint Stables'. Historic Analysis and Research Team files, English Heritage

Tucker, M 2010 'The "factory" building (former locomotive lifting shop) at Old Oak Common Railway Depot – its significance'. Hammersmith Historic Buildings Group, London

Weiler, J M 1987 'Army architects – the Royal Engineers and the development of building technology in the nineteenth century'. Unpublished Ph.D thesis, University of York Institute of Advanced Architectural Studies

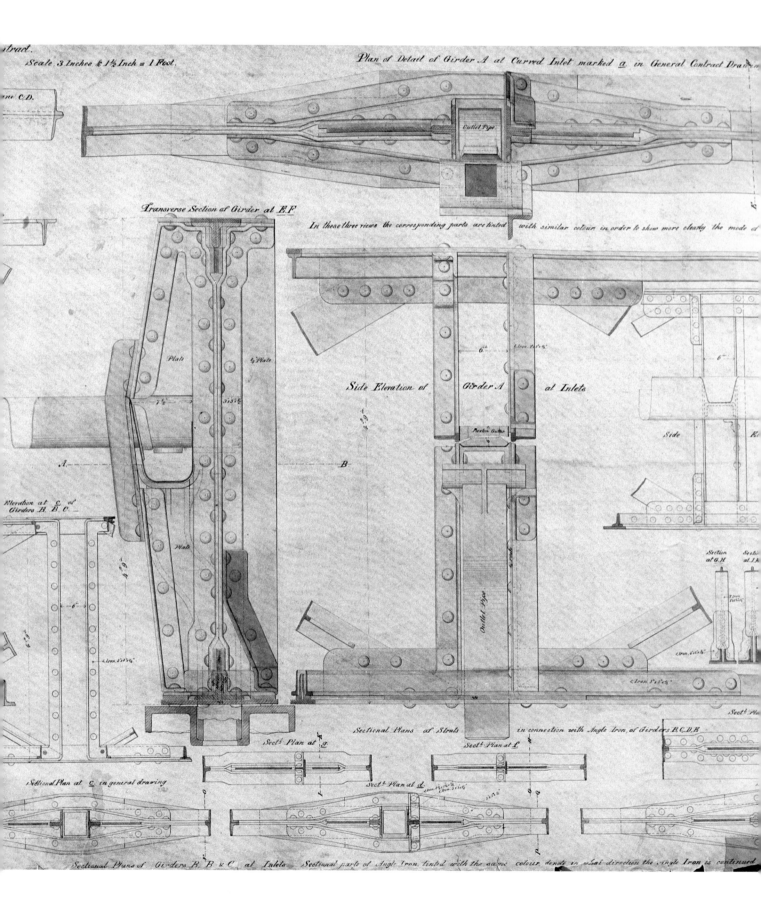

A contract drawing by Fox, Henderson & Company for wrought-iron girders in an unidentified part of the station, c 1851–4. [Network Rail/EH AA031091 (detail)]

INDEX

Page references in **bold** refer to illustrations.

A

A D Dawney & Co 65, 166n91
Acton 12, 19, 86
Alan Baxter & Associates 93
Albert, Prince Consort 9, 24, 30, 48
Andrew Handyside, Messrs (Derby) 151, 165n48
Anglo-American Electric Light Company 57
Argus (locomotive) 11
Armstrong, William (GWR engineer) 59, 139, 144, 171n56
    and Bishop's Road Bridge 151, 152, 156
    and Span Four 119, 165n67
Armstrong, Sir William, later Lord Armstrong 43, 57, 71–2, 73
Armstrong, Sir William, & Company, Elswick (on Tyne) 43, 71, **72–3**, 134
arrival side offices (1932–3) 65, **65**, 125
Atlantic trade 9
Aukett Associates 168Ch6n5
automatic ticket machines 87
automatic train control 64
Ayrton, Acton Smee 53–4

B

Baker, George, & Son, contractors 29
Bakerloo Line, platforms and concourse 93, 119, 130
Banbury 75
Bank Holidays 56, 83, 87
Barlow, W H 31
Barry, Sir Charles 30
Bath 2, 4, **5**, 8, 84
    'Skew Bridge' across Avon 103
    station 28, 34, 103
Baxter, Alan, & Associates 93
Beenham Grange, offices evacuated to 68, 118
Belfast, glasshouse 29

Bennett, Joseph 6
Berks & Hants Railway **5**, 10
Bertram, Thomas A 6, 32, 33, 37, 39, 43, 169n10
    as resident engineer 48, 50, 117
'Big Four' railway companies (formed 1923) 62, 64, 66, 69, 93, 140
Birmingham 1, 4, **5**, 10, 26, 32, 56, 77, 92
    New Street Station 39, 49, **49**, 91
    see also London & Birmingham Railway
Birmingham, Bristol & Thames Junction Railway 54
Birmingham, Wolverhampton & Dudley Railway **5**, 26
Birmingham & Oxford Junction Railway **5**, 26, 34
Bishop of London's estate 16, 57
Bishop's Road, former down parcels depot 65, 87, **100**, 145
Bishop's Road Bridge 17, 19–21, **20–1**, 23, 66, 151–2, **151–2**
    Bishop's Road Canal Bridge 20, 21, 24, 25, 152–7, **154–7**
    and Crossrail project 96
    and Paddington LTVA (Long Term Vehicle Access) Project 95, 151–2, **152**, 156–7
    spans replaced (1906–7) 25, 60, **105**, 151
Bishop's Road Station 54, **54**, 56, 59, 66, **105**, 119, 128, 130, **131**
    rebuilt (1930s) 121, 128
    see also Metropolitan and Hammersmith & City Line platforms (former Bishop's Road Station)
'Blue Pullman' DMUs 140, 142
boardroom 84, 117–18, **118**
Boer War **76**, 118
bomb damage, Second World War 68, 69, 90, 140
    Eastbourne Terrace buildings 68, **69**, 90, 111, 112, 113, 114, 118

Bond, Michael, Paddington Bear books 90, **91**
booking offices 84, 87
    booking hall (1880s) 112
    main booking hall (1850–5 station) **35**, 56, 71, 114, **114–15**
    see also ticket offices
Bourne, J C **9**, **14**, **21**, 23, **27**, 78
Box Tunnel 4, 6, 8–9, **9**, 40
Bradford-on-Avon 4
Bramah, Fox & Co 38
Bramah, Francis 38
Breguet, Louis 2
Brentford 56
Brereton, Robert P 6
bridges at Paddington 150
    moveable platform bridges (1850–5 station) **35**, 72
    see also Bishop's Road Bridge; Lord Hill's Bridge; Ranelagh Bridge; Westbourne Bridge
Bridgwater 9
Bristol
    Clifton suspension bridge 2–3, 34, 162n44
    Floating Harbour 1, 3, 9, 159n51
    and origins of Great Western Railway 1–2, 3, 4, 9, 92
    railway to
        construction 8, 9
        journey times and fares 11, 74, 79, 80
        last broad-gauge train 58
    rivalry with Liverpool for Atlantic trade 9
    Temple Meads, Bristol & Exeter station 9
    Temple Meads Station 7, 8, 27, **27**, 28, 33, 34, 39, 76, 103
        reconstruction (*c* 1865–77) 50
        remodelled (1930s) 65
        traversers at 22
Bristol & Exeter Railway 3, **5**, 9, 58
Bristol & Gloucestershire Railway 1
Bristol Dock Company 1, 2, 3
Bristol Railway 1, 3, 12
British Rail 91–2, 93, 136, 140, 141, 148, 150

British Railways Act (1947) 69
British Railways Board 69, 90, 140
British Transport Commission 16
British Waterways 157
broad gauge *see* gauge issues
Brown Boveri 140
Brunel, Isambard, Junior 3, 72, 73, 100
Brunel, Isambard Kingdom
    general
        bust and statue **84**, 111, 118
        death (1859) 50, 85
        legacy 141
        life/career/character 2–3, 34, 46–7, 50, 79
        London houses and offices 3, 6, 7, **7**
        photographs **1**, 50
        views on ornament 39–42, 44, 98, 106–8
    and Great Exhibition 30–2, 34
    and Great Western Railway 1–11
    and Paddington Station (1835–50) 12–25
        Bishop's Road Canal Bridge 152–7, **154–5**
        unexecuted designs 17–18, **17–18**, 27–8
    and Paddington Station (1850–5) 27–50, 70–3, 75–7
        designs for Conduit Street frontage 17, 27, 32, 33, 99
        designs for Eastbourne Terrace buildings 39, 109, 113, **113–14**, 114
        designs for goods station **43**, 131–2
        designs for station layout 27–8, **28**, 32–3, **33–4**, 70–1
        designs for train shed 28, **28**, 33, **33–4**, 39, 40–2, **41**, 100–8, **106**
    and underground railways 53
Brunel, Sir Marc 2
Brunton, William 2, 4
Buckingham Railway **38**, 39
Builder, The, on opening of new station (1854) 48, **107**, 108
Bulkeley (locomotive) 58
Burge, George 8

Burke, James St George 32
Burlinson, N J 58
Burnet, Tait & Lorne 108
  see also Tait, Thomas
Burton, Decimus 29, **29**, 103,
  169n64
bus services 66

C

cab access **35**, 60, 71, 76, 121
  cab road ('inclined plane') 71,
    109, **109**
  glazed roof 109, 112, 113,
    113
  see also taxi access
Camden Town 13
canals 10
  see also Grand Junction Canal;
    Grand Union Canal
Canary Wharf 96
Canterbury & Whitstable
  Railway 2
Carpmael, Raymond 65–6, 128,
  170n49
carriage sheds
  1835–50 station **20**, 22
  1850–5 station 131, 144, **144**
  see also Old Oak Common
carriage trucks see horses and
  carriages
carriages (railway) 58, 72, 73–4,
  77, 81
  royal 24
carriages (road) see horses and
  carriages
carriers 24, 78, 83, 131
Cass Hayward (engineers) 156
Cassell & Co, Official Guide to the
  Great Western Railway 82
'Castle' class locomotives 63, 139
Cavendish, William, 6th Duke of
  Devonshire 29, 30–1
Central Line 56
ceremonial events, Paddington as
  venue for **77**, 78
Chalk Farm 13, 146
Chance, Richard 31, 106
Chance Brothers,
  glassmakers 31–2, 106
Charing Cross Station 53, 60
Chatham, naval dockyard 29
Chatsworth, Joseph Paxton at 29,
  103, 106
  'Great Stove' 29, 103, 106
  Victoria Lily House 31–2, 106
Cheltenham **5**, 10, 63
Cheltenham & Great Western
  Union Railway **5**
Cheltenham Spa Express 63
Chepstow, Wye Bridge 34, 47,
  83, 162n44
Chester, Dee railway bridge 156

children as employees 84
Chippenham 4, **5**, 8, 83
Christie, Agatha, 4.50 from
  Paddington 90
Christmas
  decorations **123**
  mail services 89
Churchill, Victor Spencer, 1st
  Viscount 61
Churchward, George Jackson 58,
  63, 91, 137–9, 140, **141**, 142
Circle Line 56
  see also Praed Street Station
    (now Circle Line platforms)
'City' class locomotives 58
Clapham Junction 55
Clark, C W 130
Clark, George T 6
Clarke, Seymour 24
class distinctions 78–81, 85
  see also first class; second class;
    third class
Cleveland Bridge & Engineering
  Co 135, 165n70, 166n93
cloakroom and left luggage office
  (1850–5 station) **35**, 75
clock arch see entrance loggia
  (later clock arch)
coal depot see goods station
coal trade 135
  Welsh 58, 61, 62, 83, 131, 132
Cockerell, Charles Robert 30
Cockerell, John 16
Cockerell, Samuel Pepys 16
Cole, Sir Henry 30, 31
Collett, C B 63, 91, 139, 142
colour light signalling 64
Colthurst, J 154
columns, in train shed 40–2,
  60–1, 92, 102, 103–4, 106
commuter services 53, 93
  see also suburban lines and
    services; underground
    railways
Conder, F R 8, 158n20, 159n38
Conduit Street (now Praed Street)
  Brunel's plans for station
    frontage 17, 27, 32, 33, 99
  see also Praed Street Station
Connell, George 92
Cooke, William Fothergill 24
Cooper, ____ 84
Cornish Riviera service 61, 66
Cornishman Express 58, **59**
Cornwall Railway **5**, 10, 46, 58
corridor trains 81
Coulsell, Mary 24
Cragside 43, 57
cranes
  1835–50 station 43, 71
  Old Oak Common, overhead
    travelling cranes **138**, 141,
    142, **143**

crescent (sickle) trusses 29, 39,
  102, 103
Crimean War, prefabricated
  hospital building for 50
Cromford & High Peak
  Railway 158n8
Crompton dynamos 57
Crossrail Act (2007) 96
Crossrail project 93, 95–6,
  97–100, 113, **113**, 122, 136
  depot at Old Oak Common 137,
    140
Crutched Friars, stables 83
Crystal Palace 26, 30–2, **30**, 39,
  96, 100–2, 103
  Owen Jones colour scheme 37,
    108, 163n85
  Paxton glazing (Paxton
    roofing) 31, 40, 106
  at Sydenham **44**, 49, **86**
  see also Great Exhibition (1851)
Cubitt, Lewis 14, 15, 27, 103
Cubitt, William 14, 15, 30
Culverhouse, P A 65–6, 112, 117,
  128, 145
  arrival side offices 65, **65**, 125
  Eastbourne Terrace offices 65,
    112
  Great Western Royal Hotel 65,
    66, 127, 166n91
  Lawn, the 65, 94, 123, **125**
  refreshment room 114, **115**,
    118
  staff hostel 145

D

Dawney, A D, & Co 65, 166n91
day trips see excursion trains and
  day trips
de Maré, Eric **120**
Dean, ____ 57
decoration see ornament
Depression (1929–31) 64, 66
'descending platforms' 72, 73, 77,
  100
Dethier, Monsieur 44–6
Development (Loans, Guarantees
  and Grants) Act (1929) 64
Devizes 2, 4
Devonshire, William Cavendish,
  6th Duke of 29, 30–1
Didcot 4, **5**, 8, 84
  Great Western Trust 136
diesel locomotives 140
diesel multiple units (DMUs) 140,
  142
diesel railcars 64
dining cars 81
directors' smoking room 118,
  118
discipline, industrial 83–5
Dobson, John 29, **29**, 102, 103

dockyards, naval, iron
  structures 38, 39
Donaldson, T L 30
Dowbiggin, Holland & Co 44
Dublin, Turner's foundry 29
Duquesney, François 27

E

Ealing 8, 15, 53, 56, 146
Eastbourne Terrace
  buildings **99–100**, 108–18,
    109–18
  construction (1850–5) 33, **35**,
    37, 39
  and Crossrail project 95, 96,
    97–8
  later alterations 56–7, **57**, 65,
    90
  war damage 68, **69**
Eastern Counties Railway 158n19
Eastern Steam Navigation
  Company 34, 162n44
Edmondson, Thomas,
  'Edmondson ticket'
  system 71, **71**, **87**
Edward VII, King
  funeral 78
  as Prince of Wales 48
Egham, Royal Holloway
  College 52, **52**
Electric Telegraph Company 24
electricity
  electric light installed
    (1880s) 56–7, **58**
  electricity substation,
    Porchester Road 60, 144,
    145
  electrification of Metropolitan
    Railway 56, 60, 145,
    164n36
  generating station, Park
    Royal 57, 139, 144
  overhead electrification
    (1980s) 94
Elkington Shield 86, 118
Ellis, John 31
Elswick (on Tyne) see Armstrong,
  Sir William, & Company
employees see staff
engine sheds 136, 137
  1835–50 station, roundhouse
    (1837–8) **20**, 22, **22**, **25**,
    136
  Old Oak Common 137, 138–9,
    138–9, 140
  Westbourne Park 136
engineering department 50, 60,
  150
English Heritage, and Bishop's
  Road Canal Bridge 156–7

entrance loggia (later clock arch) **35**, 71, 92, **99**, 107, 111, 114
escalators 130
Euston Station 14–15, **14–15**, 18, 26, 33, 38, 62
  hotels 32, 34, 44, 126
  as proposed joint terminus for GWR and London & Birmingham Railway 12–15
  rebuilding (1960s) 91
excursion trains and day trips 56, 66, **67**, 81, 82–3, 87
Exeter 3, **5**, 9, 10, 58

F

Fairbairn, Sir William 103
fares *see* tickets and fares
Farringdon Station 54, 128, 130
'Festival of Britain' style 90, **90**
'Firefly' class locomotives 11, **11**
first class 71, 73–4, 78, 79, 80, 81, 82
  waiting rooms (1850–5 station) **35**, 56, 71, 75, 107, 111, 114
First World War 61–2, **61**
Flachat, Eugène 29
Flatow, Louis 51
Fleetwood, North Euston hotel 44, 169n64
Fleetwood-Hesketh, Sir Peter 169n64
Flying Dutchman express service 80
Forces' canteen (1919) **61**
fourth span *see* Span Four
Fowler, Charles, Junior 40
Fowler, Sir John 128
Fox, Sir Charles 31, 37–9, **37**, 49, 50
  and Paddington Station (1850–5) 37, 43, 46, 49, 100, 102
  see also Fox, Henderson & Co
Fox, Sir Charles, & Sons 49
Fox, Francis 50
Fox, Henderson & Co 38–9, 49, 94
  and Crystal Palace 29, 31–2, 49
  and Paddington Station 37–9, **38–9**, 42–3, 46–7, 102–6, **104**, **176**
Freeman Fox & Partners 49
freight *see* goods traffic
Frere, G E 6
'Friday night specials' 83
Friend, ____ 167n51
Frith, William Powell, *The Railway Station* 51–2, **52**, 70, 118
Fry, Samuel 52
Fryer, ____ 84

G

Gabowen 84
Gainsford, Charles 39, 43, 46, 168Ch7n11, 169n10
Galvanised Iron Company 34, 162n44
gas-turbine locomotives 140
Gauge Act (1846) 10
gauge issues 4–6, 10, 13, 15, 52–3, 56, 58, 136
  mixed-gauge track 10, **10**, 52, 53, 54, 58
General Strike (1926) 63, **63–4**, 64, 86
George V, King, funeral 78, **79**
George VI, King 52
  funeral 78
George Baker & Son, contractors 29
Gibbs, ____ 24
Gladstone, W E 80
glasshouses (greenhouses) 29–31, **29**
Gleadow & Armstrong 165n67–8
Glennie, William 6
Gloucester **5**, 10
Gloucester & Forest of Dean Railway 5
Gooch, Sir Daniel **48**, 50, 51, **53**, 91
  as GWR chairman 51, 52–3, 57–8
  as locomotive superintendent 6, 11, 24, 52, 131
  engine house (1837–8) 22, **22**, 136
  office at Westbourne Park 136, **137**
'goods passengers' 78–9
goods station 131–6
  1835–50 station 17, 19, **20**, 22, **23**, 32, 42, 131
  1850–5 station 42, **43**, 46, 83, 84, 131–4, **131–5**
  goods offices 60, **131**, 134, **135**, 136
  granary **43**, **131**, 132, **133**
  high-level goods yard/coal depot and viaduct **35**, 60, 83, 119, 121, 128, **131–3**, 132–4, **135**
  goods offices (c 1906–7) 60
  rebuilding (1925–6) 62, **62–3**, 134–5, **135**
  redevelopment of site **100**, 135–6
goods traffic 24, 78–9, 83, 87–8, 131–6, 148
  see also goods station; road transport
Gordon, Messrs (Deptford) 154, 155
Goring 83

Goswell Road stables 146
Gough, A D 108
Graham, ____ 84
granary *see* goods station
Grand Junction Canal 16, 17, 103, 119, 131–2, **131**, 153
  Bishop's Road Canal Bridge **20**, 21, 24, 25, 152–7, **154–7**
  Paddington Basin 16, 119
Grand Junction Railway 1, 4, 10, 158n19
Grand Union Canal 138
Grant, ____ 167n51
Grant, Sir Francis 51, **51**
Graves, Henry 51
Great Britain (locomotive) 52
Great Britain, SS 9
Great Central Railway 56
Great Depression (1929–31) 64, 66
Great Eastern, SS 34, 47, 50, **50**
Great Eastern Railway 53
Great Exhibition (1851) 26, 30–2, **30–1**, 34, 37, 40, 161–2n31, 163n85
  see also Crystal Palace
Great Northern Railway 27
Great Western (locomotive) 58, **59**
Great Western, SS 9
Great Western Railway (GWR)
  history
    origins 1–11
    named (1833) 3
    opening (1838) 19
    and Railways Act (1921) 62
    centenary (1935) 66
    nationalisation (1948) 69
  Acts of Parliament
    Great Western Railway Bill, first (1834) 3–4, 5, 12
    Great Western Railway Act (1835) 4, 5, 13
    Great Western Railway Extension Act (1837) 16, 18
  'God's Wonderful Railway' or 'the Great Way Round' 58, 98
  routes 4, 5, 92–3
  track design 4–6
  see also gauge issues; suburban lines and services
Great Western Railway Literary Society 85–6
  Conversazione (1859) 85–6, **85**
Great Western Railway Magazine 86
Great Western Royal Hotel 32, 33–4, **33**, **35–6**, **99–100**, 126–7, **126–7**
  construction 42, 44–6, 48

extension and refurbishment (1930s) 65, 66, 127
  track to take luggage to station **35**, 75
Great Western Steamship Company 9
Green, F W 87
greenhouses *see* glasshouses
Gresley, Sir Nigel 64
Grierson, ____ 152, 156
Griffiths, Henry 84
Griffiths, Joseph 161n1
Grimshaw, Nicholas 104
  see also Nicholas Grimshaw & Partners
Grissell & Peto, contractors 15, 18–19, 150, 159n39
Gutch, George 16
GWR *see* Great Western Railway

H

Haigh Foundry 10
Hamilton, Paul 148
Hammersmith 54, 55, 56
Hammersmith & City Railway (later Hammersmith & City Line) 54, 164n36
  see also Metropolitan and Hammersmith & City Line platforms (former Bishop's Road Station)
Hammond, John W 3, 6, 15
Handyside, Messrs Andrew (Derby) 151, 165n48
Hanwell 8, 15
Harbury 84
Hardwick, Philip 14, 15, **15**, 18, 32, 33, 161n48
Hardwick, Philip Charles 26, 32, 34
  Great Western Royal Hotel 33–4, **36**, 44, 66, 126–7, **126–7**
Harrow Road, motor vehicle maintenance depot **100**, 148–50, **149**
Hawks Crawshay, contractors **29**, 102, 103
Hawkshaw, Sir John 60
Hayes, canal bridge 153
'head plan' 27
Heathrow Express 93–5, **125**
Henderson, John 38
  see also Fox, Henderson & Co
Henley 53, 82, 83
  excursion trains for Henley Regatta 66, **67**
Hennebique, François 60
Hennebique reinforced concrete system 60, 119, 144, 147
Heywood, S H, & Sons (Reddish) 141
Higgs & Hill 136

'High Paddington' 90–1, **91**, 135
High Wycombe 53, 55, 56, 137–8
Hill, General Sir Rowland, 1st
    Viscount 16, 17, 150
Hilton Paddington Hotel 127
    see also Great Western Royal
    Hotel
Hitchcock, Henry-Russell 42, 91,
    102, 106–7, 108, 126
Hobbs, George 83–4
Hodgkinson, Eaton 153, 154, 156
holiday train services 56, 58–9,
    61, 66–7, 81–3, **81**
    see also excursion trains and day
    trips
Holl, Francis 51
Holland, ____ (Grand Junction
    Canal Company) 153
Holland, Messrs 42, 44, 46, 47,
    48, 116, 126
Holliday & Greenwood,
    Messrs 60, 65, 119, 165n61
Holloway, Thomas 52
Hopkins, Michael, & Partners 93
Horeau, Hector 30
'horse arch' 75, **99**, 112, 166n15
horse-drawn tramways 4, 158n8
Horseley Bridge & Engineering
    Company 60, 119, 150
horses, use by GWR 62, **82**, 83,
    88, 146–8, **147–8**
    see also Mint stables
horses and carriages, carried on
    railway 17, 18, 19, 21, **27**,
    35, 74–5, 76
hostels see staff hostels
hotel see Great Western Royal
    Hotel
Huckle, ____ 84
Hudson, George 26
Hungerford 2, 4, **5**, 53
'hunting services' 75
Hyde Park see Crystal Palace
Hyder plc 49
hydraulic machinery (1850–5
    station) 71–3, 134

I

immigrants, West Indian **90**
'inclined plane' see cab access
Indented Bar & Concrete
    Engineering Company 134
industrial discipline 83–5
'Iron Duke' class locomotives 52
island platforms (1850–5
    station) **35**, 71

J

Jackaman, Messrs (Slough) 60,
    151, 165n48, 165n60,
    165n62–3, 165n66

Jacomb, W 39
Jagger, Charles Sargeant 61, 108,
    108
John McAslan & Partners 94
John Scott-Russell & Company 50
Jones, ____ 84
Jones, Owen 37, 44, **44**, 98, 108

K

Kadleigh, Sergei 90, 135
Keay, E C & J 165n85, 165n88,
    171n64
Kew Gardens 31
    Palm House 29, **29**, 30, 103
Killingworth Colliery,
    tramway 158n22
'King' class locomotives 63, 139
King's Cross Station 27, 103
Kinnaird, ____ 86
Kirk & Randall (Woolwich) 56,
    112, 164n27, 168n30,
    168n33, 168n38

L

L G Mouchel & Partners 60,
    165n53, 171n56
La France (locomotive) **141**
Lamb, E B 107, 108
Lambert, Jasper 164n8
lamp room (1850–5 station) **35**,
    71
Latimer, John 7–8
Lawn, the 46, 56, 87, 94, **94**,
    99–100, 123, **123–5**
    glass roof (1933) 65, **66**, 123
    glass roof (1990s) 94, 123, **124**
left luggage office (1850–5
    station) **35**, 75
libraries 85
Liverpool
    Crown Street Station 27
    Lime Street Station 27, 29, 103
    'Liverpool party' 6, 9, 13–15
    Mather, Dixon & Co 10
    rivalry with Bristol 1, 4, 9, 92
Liverpool & Manchester
    Railway 1, 2, 4, 24, 38
Liverpool Street Station 53, 96
Lloyd, J C 170n16
Locke, Joseph 29, 103, 158n3,
    169n64
Locke & Nesham, Messrs 46, 136
locomotive department
    Old Oak Common 60, 87, 137–
    44, **138–9**, **141–3**
    Westbourne Park 76, 136–7,
    136–7
locomotive superintendent's
    house, new station
    (1850–5) 48, **48**

locomotives 10–11, 58, **59**
    alternatives to steam
    power 140, 142
    'Castle' and 'King' classes 63,
    79, 139
    'City' and 'Saint' classes 58
    'Iron Duke' class 52
    'Star' and 'Firefly' classes 11,
    11, 58, 159n57
    North Star 6, **6**, 11, 19
    see also locomotive department;
    individual locomotives by
    name
loggia see entrance loggia
London
    18 Duke Street, Brunel's
    office 6, 7, **7**
    53 Parliament Street, Brunel's
    house 3
    Bishop of London's estate 16,
    57
    transport network 93
    see also suburban lines and
    services; underground
    railways; *individual*
    buildings and locations by
    name
London, Brighton & South Coast
    Railway 53, 55
London, Chatham & Dover
    Railway 52, 53
London, Midland & Scottish
    Railway 165n73
London & Birmingham Railway
    (L&BR) 1, 4, 6, 11, 24, 92,
    153, 158n19
    and Euston Station 12–15, 38
    gauge 4, 5
    merger with Grand Junction
    Railway 10
London & Bristol Railway 12
London & North Eastern
    Railway 64, 165n73
London & North Western
    Railway 32, 58, 80, 91, 92–3,
    146, 169n64
    formed by merger of London &
    Birmingham Railway and
    Grand Junction Railway 10
    suburban services 53, 54–5
    see also Birmingham, New
    Street Station; Liverpool,
    Lime Street Station
London & South Western
    Railway 56, 77
London & Southampton
    Railway 158n19
London Bridge Station 53
London Street
    diversion 59, 60, 119
    London Street Deck 119–22
    sorting office 89
London Underground Limited 95
    see also underground railways

'Long Charlies' (carriages) 72, 77
Lord Hill's Bridge 17, 22, 60, 150
Lord's Day Observance Society 82
lorry depot, Westbourne Park 63,
    121
Lucca, Duke of 24
luggage
    left luggage office (1850–5
    station) **35**, 75
    luggage bins 76
    track to take luggage from hotel
    to station **35**, 75

M

McAslan, John, & Partners 94
MacDermot, E T 58
McIntosh, Hugh and David,
    contractors 8, 159n39
Macmillan House 118
Maidenhead 8, 18, 19, 22, 24
    bridge 8, **8**, 19
    station 22
mail services 65, 75–6, 88–9, 93
Manchester 4
    Liverpool Road Station **2**
Mansion House Station 56
market traffic 87–8
Marlow 53
Marsh, T E 6
Marylebone Station 62
Mather, Dixon & Co
    (Liverpool) 10
Matthews, E C 71, 73, 75, 76
Maudslay, Son & Field 2
Measom's *Illustrated Guide to the*
    Great Western Railway 82
Menai Straits, Britannia
    Bridge 31
Metropolitan District Railway 56
Metropolitan and Hammersmith
    & City Line platforms (former
    Bishop's Road Station) 66,
    93, 94, **99–100**, 121, **122**,
    128, **129**
    and Crossrail project 95, 122
Metropolitan Railway 53–4, 55,
    55, 93, 134
    electrification 56, 60, 145,
    164n36
    see also Bishop's Road Station;
    Praed Street Station
Michael Hopkins & Partners 93
'Middle Circle' service 56
Midland Railway 10, 31, 37, 80,
    81, 83, 93
milk platform **35**, 56, 59, 60, 75,
    111, **112**, 117, **117**
    later replacements 119, 144
milk traffic 83, 87, 88, **130**, 135
    see also milk platform

Mint stables (now Mint Wing,
St Mary's Hospital) 60, 63,
82, 83, 88, **99–100**, 119,
146–8, **146–8**
mixed-gauge track 10, **10**, 52, 53,
54, 58
see also gauge issues
Monmouth & Hereford
Railway 10
Moorgate Station 54, 56
Morning Star (locomotive)
159n57
Morton, William Scott 52
motor transport *see* road transport
motor vehicle maintenance depot,
Harrow Road **100**, 148–50,
149
Mott MacDonald 122
Mouchel, Louis-Gustave (L G
Mouchel & Partners) 60,
165n53, 171n56
moveable platform bridges
(1850–5 station) **35**, **72**

N

nationalisation (1948) 69, 140
naval dockyards, iron
structures 29, 38, 39
navvies 8–9
Network Rail plc 95, 145, 151
New Orleans Railway 11, 159n57
Newcastle upon Tyne
Central Station 29, **29**, 31, 39,
102, 103
Robert Stephenson & Co 11
newspaper handling 88, 91
Nicholas, J S 165n88, 165n90
Nicholas Grimshaw &
Partners 93–4, 95, 123, **124**
'Night Goods' passenger
service 80
North Star (locomotive) 6, **6**, 11,
19

O

offices *see* arrival side offices;
Eastbourne Terrace buildings;
goods station
Olander, Edmund 56
'E O' ?identified as 168n30,
169n38
Olander, J 164n22
Old Oak Common, locomotive
department 60, 87, 137–44,
138–9, **141–3**
carriage sheds and shops 87,
138, **138–9**, 142, **143**
Crossrail depot 137, 140
engine shed 137, 138–9, **138–**
9, 140

locomotive lifting shop
('factory') **138**, 139, 140–
4, **141–3**
ornament, in 1850s station
39–42, 44, 98, 106–8
Overend, Gurney & Company,
bank crash (1866) 52, 56
Owen, Lancaster 56, 146, **146**,
168n28, 168n30, 168n38
Owings, Nathaniel Alexander 66
Oxford **5**, 10, 32, 53, 67, 75
Rewley Road Station **38**, 39
see also Birmingham & Oxford
Junction Railway
Oxford, Worcester &
Wolverhampton Railway **5**,
10
Oxford & Rugby Railway **5**, 10,
26
Oxford & Worcester Railway 34,
162n44

P

Paddington Basin 16, 119
Paddington Bear 90, **91**
Paddington Integrated
Project 95, **96**, 122, 128, **129**
Paddington LTVA (Long Term
Vehicle Access) Project 95,
151–2, **152**, 156–7
Paddington Parish Vestry 16–17,
23–4
Paddington Station, structural
development
site 16–17, **16**, 53, 92
first station (1835–50) 12–25
plan **20**
new station (1850–5) 26–50
plans **35**, **131**
expansion and development
(1855–1947) 51–69
map (1869) **54**
decline, revival and the future
(1948–) 90–8
plans **99–100**
see also individual buildings,
structures and areas by
name
parcel services 75–6, 83, 87,
88–9, 131
down parcels depot, Bishop's
Road 65, 87, **100**, 145
Red Star parcels depot 92, 94,
121–2
up parcels depot, Mint
stables 60, 119, 147
Paris
Entrepôt des Marais 29
Gare de l'Est 27
Gare du Nord 27
Park Royal, generating
station 57, 139, 144

Parker, Sir Peter 91
'parliamentary trains' 80, 81
'passimeter office' 87
patent Victoria stone 60, 66, 94,
112, 123, 144
Great Western Royal Hotel 65,
127
Paxton, Joseph 29, 30–2, 37, 49,
103, 106
Paxton glazing (Paxton roofing)
Crystal Palace 31, 40, 106
goods station 134
Lawn, the 123
shed roofs 40, 75, **99**, 100,
104–6, **105**, 113
removal 90, 91, 92, 94
surviving area **99**, 111–12
Peckham, John 83–4
Pembroke, naval dockyard 29,
38, 39
Penzance **5**, 58, **59**, 82
Persia, Shah of **77**
Peto, Sir Samuel Morton 32
see also Grissell & Peto
Peto & Betts, contractors 159n39
'Picture, the' 51, **51**, 83, **84**, 118
plan store 117, **117**
platform catering 81, **81**
Plymouth **5**, 10, 58, 80
Millbay Docks 166n17
Pole, Sir Felix C 62, 63
Polonceau truss 142, 144
Poplar Dock, goods depot 134
Porchester Road, former GWR
stationery store 60, 144–5
porters 52, 76, 83, 84, 86
Portsmouth, naval dockyard 29,
39
Post Office 65, 93
underground railway 89
see also mail services
posters 59, 66
Praed Street, Brunel's plans for
station frontage 17, 27, 32,
33, 99
Praed Street Station (now Circle
Line platforms) 54–**5**, 55, 56,
93, **100**, 128–30, **129**
Price, Henry 2, 4
printing department 71, 117, **117**
privatisation (1996) 93, 149
prize fights, special trains for 82

Q

Quainton, Buckinghamshire
Railway Centre 39

R

Railtrack plc 93–4, 95
railway accidents 79–80

Railway Regulation Act
(1844) 80
Railways Act (1921) 62
Ranelagh Bridge 17, 60, 87, 145,
150, **150**
Ranger, William 8
Ransome & Rapier (Ipswich) 121,
138
Reading 2, 4, **5**, 8, 53, 55
station
engine house 161n48
traversers 22
Red Star parcels depot 92, 94,
121–2
refreshments
refreshment rooms and
stalls **35**, 66, 71, **89**, 114,
115, 118, 125
see also dining cars; Forces'
canteen; platform catering;
restaurant
Regalian Properties plc 136
Remon, Herr 126
Rendle's patent glazing 121
Renkioi (Turkey), prefabricated
hospital building 50
restaurant **92**
see also refreshments
Reynaud, Léonce 27
road transport
for goods traffic 62–3, 66, 88,
146–8
lorry depot, Westbourne
Park 63, 121
motor vehicle maintenance
depot, Harrow Road **100**,
148–50, **149**
road transport lobby
(1930s) 66
see also cab access; carriers;
horses; Paddington LTVA
(Long Term Vehicle Access)
Project; taxi access
Roch, Nicholas 2
rolling stock *see* carriages;
locomotives
roofs *see* Span Four; train shed
Rosehaugh Stanhope 136
Roumieu, R L 108
roundhouse (engine shed), 1835–
50 station **20**, 22, **22**, **25**,
136
Royal Docks (London), goods
depot 134
Royal Holloway College 52, **52**
Royal Oak underground
station **150**
royal patronage 24, 77–8
Royal Society *see* Great Exhibition
royal waiting rooms
arrival side 75, 77
departure side **35**, 44, **45**, 47,
68, 71, 77, 92, 114–16, **116**
entrance 107, 108, 111, 113

Ruabon Coal Company 132, **132**
Ruskin, John 40
Russell, Charles 51
    portrait ('the Picture') 51, **51**,
        83, **84**, 118

S

'Saint' class locomotives 58
St Mary's Hospital, Mint
        Wing **100**, 146, 148
    see also Mint stables
St Pancras Station 62
Sale, R C 153
Saltash, Royal Albert Bridge 46,
        47, 50
Salter, Messrs 83
Sargent, R G 166n91
Saunders, Charles Alexander 3,
        3, 11, 14, 15, 24, 34, 51
    and construction of new
        station 43, 44, 46, 47, 109
Scott-Russell, John, &
        Company 50
season tickets 53
second class 71, 73–4, 78, 79,
        80–1, 82
    waiting rooms (1850–5
        station) **35**, 48, 71, 114
Second World War 68–9, **68**, 90,
        118, 140
    see also bomb damage
Severn Tunnel 56, 58
Seymour Harris & Partners 91
Sheldon Square 136
Sherwood, Messrs B & N,
        contractors 19, 42, 46, 132,
        151, 154, 155
sickle trusses *see* crescent (sickle)
        trusses
signal boxes 87, 130–1, **130**, 145
signalling, colour light
        signalling 64
Sims, ____ 24
Sitwell, Sir Osbert 68
Skidmore, Louis 66
Skidmore Owings Merrill,
        architects 66
Slough 19, 24, 60, 77, 78
Smethwick 31, 38, 103, 106
    see also Fox, Henderson & Co
Smithfield, goods depot and
        stables 83, 134
Sonning Hill 4, 79–80
South Devon Railway 3, **5**, 10, 58
South Eastern & Chatham
        Railway 61
South Kensington Station 55, 56
South Lambeth, goods depot 134
South Wales Railway **5**, 10, 34,
        46, 132, 162n44
South Western Railway 53, 55
South Wharf Road *see* Mint stables

Southall 56, 64
    canal bridge 153
Southern Railway 165n73
Spagnoletti, ____ 57
Span Four (fourth span) 59, **59**,
        60–1, **61**, 94, **100**, 119–22,
        120–2, **130**
    restoration 95, **95**, 96, **101**,
        122, **172**
'special trains' 24, 82
    see also excursion trains and day
        trips
stables 83, 146
    see also Mint stables
staff (employees) 24, 83–6
    cultural and social
        activities 85–6
    see also porters
staff hostels
    1850–5 station **35**, 48, 145
    Westbourne Terrace 66, **100**,
        145, **145**
Staines 53, 77, 86
'Star' class locomotives 11, 58
'stars and planets' motif in train
        shed arches 103, 119
stationery stores
    Porchester Road (1906–7) 60,
        144–5
    Westbourne Terrace (1932–
        5) 66, 145, **145**
Stephenson, George 4
Stephenson, Robert 4, 7, 38, 85,
        98
    bridges 31, 153, 154, 156
    and Euston Station 13, 14–15
    and Great Exhibition 30
Stephenson, Robert, & Co 11
Stevens, Alfred 44, **45**, 116
Steventon 8, 22
Stockton & Darlington Railway 2,
        4
Stone, Marcus 52
'stressed skin' design 104
suburban lines and services 53–6,
        93, 128
    see also underground railways
subway (1850–5 station) **35**, 71
Sultan (locomotive) 52
Sunderland Docks Company 34,
        162n44
Surrey Iron Railway 158n8
Swan, Joseph 57
Swansea 5, 10, 60
Swindon **5**
    GWR engine depot and
        workshops 11, 38, 137,
        142, **143**, 147, 161n51
    Mechanics Institute 85
    plan store 112, 117
    Swindon GWR band **86**
    train services 11, 58, 63, 80
    Western Region
        headquarters 92

Sydenham, Crystal Palace **44**, 49,
        86

T

Tait, Thomas 61, 108
Tanner, Sir Henry 64, **65**
Tapper, William 168Ch7n11
taxi access 87, 94–5, **96**, **99–100**,
        113, 121, **122**, 151
    see also cab access
Telegraph Construction &
        Maintenance Company 57
telegraph system 24
telephone system 87, **88**
Thames Tunnel Company 2
third class 71, 74, 78–80, 81
Thomas, John 126, **127**
Thompson, Francis 22
ticket offices 91, 92
    see also booking offices
tickets and fares 24, 71, **71**, 79,
        82–3, 87, **87**
Tim (collecting dog) 86
timetables 78, **80**
Todd, T S 64
Torbay Limited **81**
tourism 81–3
    see also holiday train services
Townsend, William 1, 3
Traffic Committee 22, 59
train shed
    architecture **99–102**, 100–8,
        104–8
    planning and construction
        34–44, **39**, **42**, 46–8
        Brunel's sketches 28, **28**, 33,
            33–4, **41**, **106**
    replacement of columns (1922–
        3) 60–1
    restoration (1985–93) 92, 103,
        106
    see also Paxton glazing (Paxton
        roofing)
transepts **70**, 72–3, 100–2, 107
Transport for London 95
traversers (traversing frames)
    1835–50 station **20**, 22
    1850–5 station **35**, 71, 72–3,
        73–4, 100
    Old Oak Common locomotive
        department 141
Tredgold, Thomas 153
Trevithick, Francis 171n82
Trowbridge 2, 4
Tucker, Malcolm 156
Turner, J M W **8**
Turner, Richard 29, 30, 103
turntables
    1835–50 station 21, 136
    1850–5 station 33, **35**, 43, 56,
        70, 71, **72**, 75, 76–7, 132

Old Oak Common locomotive
        department 137, 138, 139,
        140
    Westbourne Park locomotive
        department 136–7
Twyford 8
Tyrell, ____ 167n51

U

underground railways 53–4, 55,
        56, 95
    see also Bakerloo Line;
        Metropolitan and
        Hammersmith & City Line
        platforms; Metropolitan
        Railway
Uxbridge 53
Uxbridge Road Bridge 153, 154,
        156

V

Vale of Neath Railway 34,
        162n44
Vantini, ____ 44
vehicle maintenance department
    see motor vehicle
        maintenance depot
Victoria, Queen 24, 47, 77–8
    funeral 78, **78**
Victoria Station 55, 61
Victoria stone *see* patent Victoria
        stone
Vulcan Foundry (Warrington) 10
Vulcanized Rubber Company 34,
        162n44

W

wagon sheds 144
waiting rooms 66, 75
    first-class **35**, 56, 71, 75, 107,
        111, 114
    second-class **35**, 48, 71, 114
    see also royal waiting rooms
Wales, coal trade 58, 61, 62, 83,
        131, 132
Walkerdine, William, Messrs
        (Derby) 138
war memorial (platform 1) 61,
        90, 108, **108**
Warrington 1, 10
wars *see* Boer War; First World
        War; Second World War
Waterloo Station 33, 53, 67, 93
    former Eurostar terminal 94
Webb, ____ 24
West Cornwall Railway **5**, 34, 58,
        162n44
West Ealing, 'horse hospital' 146

West London Junction
	Railway 137, 138
West London Railway 54
Westbourne Bridge 17, **25**, 60,
	150–1
Westbourne Park
	goods depot 134
	locomotive department 76,
		136–7, **136–7**
	lorry depot 63, 121
Westbourne Park Station 54,
	136–7
Westbourne Place 16, 150
Westbourne Terrace, former GWR
	stationery store and staff
	hostel 66, **100**, 145, **145**

Western Region 90–2
Westminster City Council 93,
	151, 156, 157
Weston Williamson 122
Westway flyover 93, 136, 148,
	149
Westwood & Company 151
Wharncliffe Viaduct 8, 15
Wheatstone, Charles 24
Widows and Orphans Fund 86
Wild, Charles Hurd 49
Wilder, _____ 43
Willans engines 57
William Walkerdine, Messrs
	(Derby) 138

Wiltshire, Somerset & Weymouth
	Railway **5**, 10
Windsor 24, 53, 77–8, 82
Windsor Castle (locomotive) **79**
Winsland Street *see* Mint stables
Wodehouse, P G, *Uncle Fred in the
	Springtime* 66–7
Wolverhampton **5**, 10, 52
Wood, Nicholas 158n24
Wood Lane, milk depot 88
Woodstock 75
Woolaston, _____ 164n8
Woolwich, naval dockyard 38
Wootton Bassett 4, 82
World Wars *see* First World War;
	Second World War

Wyatt, Sir Matthew Digby **37**,
	49–50
	and Paddington Station 34–7,
		39–42, **42**, 43–4, 48, 98,
		102, 103–4, 106–8, **107**
	Eastbourne Terrace
		façade 109, **111**
	royal waiting room 44, 116
Wyatt, Thomas Henry 40

Z

Zealand Railway (Denmark) 49
Zulu express service 80